D0070595

CURIOUS CURVES

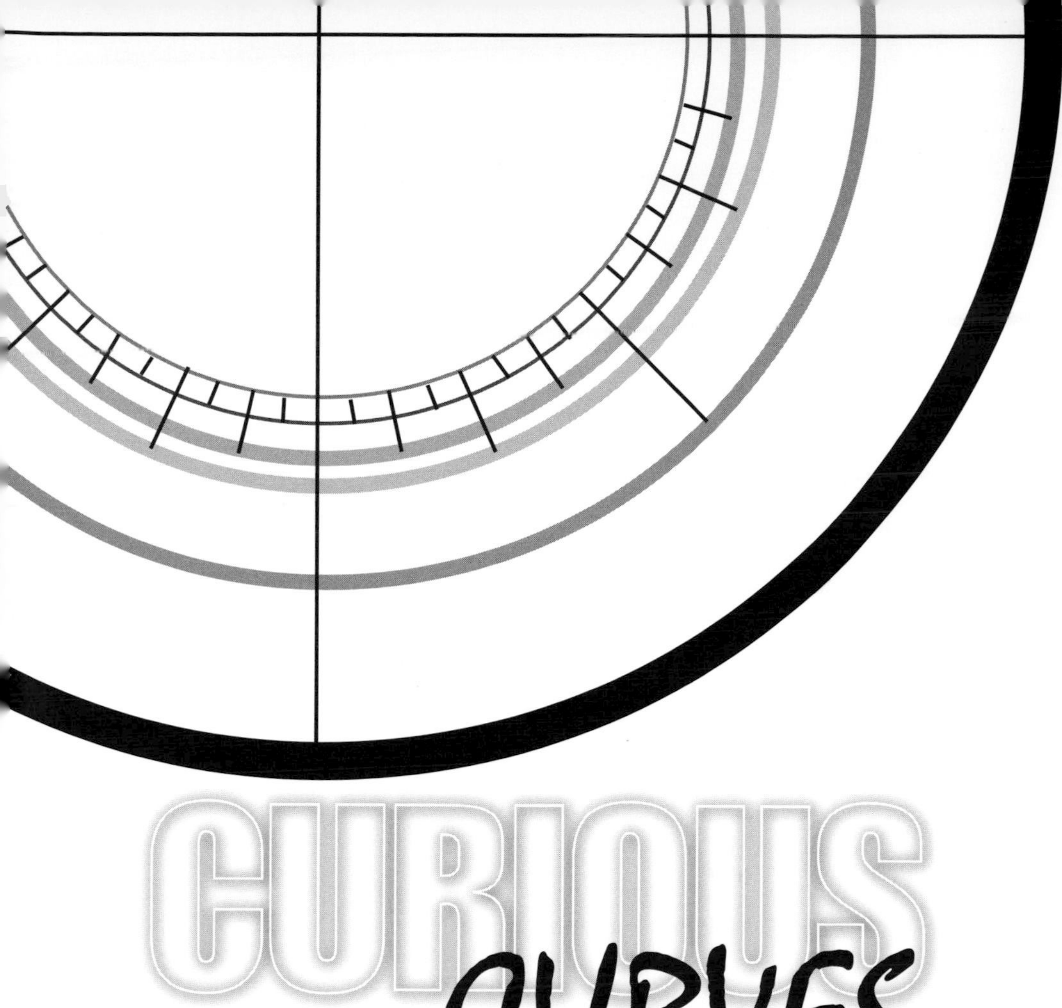

CURIOUS CURVES

Richard B Darst
Colorado State University, USA

Judith A Palagallo
The University of Akron, USA

Thomas E Price
The University of Akron, USA

NEW JERSEY • LONDON • SINGAPORE • BEIJING • SHANGHAI • HONG KONG • TAIPEI • CHENNAI

Published by

World Scientific Publishing Co. Pte. Ltd.
5 Toh Tuck Link, Singapore 596224
USA office: 27 Warren Street, Suite 401-402, Hackensack, NJ 07601
UK office: 57 Shelton Street, Covent Garden, London WC2H 9HE

British Library Cataloguing-in-Publication Data
A catalogue record for this book is available from the British Library.

CURIOUS CURVES

ISBN-13 978-981-4291-28-6
ISBN-10 981-4291-28-5

Printed by FuIsland Offset Printing (S) Pte Ltd, Singapore

Preface

This book explores captivating curves and related sets in the plane. Cantor sets, curves, maps, fixed points, and the mathematics of fractal geometry interact to produce remarkable examples. We present a sampling of classical results and introduce some fascinating new curves. We consider the curves as paths in the plane, examine visual images of them, and determine their topological properties. The material is presented with emphasis on the geometric intuition characteristic of the study of curves. Interest in curves has not diminished over the years. The advent of computing power to produce visual images has served to make the study of curves and their properties even more intriguing. Many examples and illustrative figures accompany our discussion of curves. We intend that the surprising features of these curves will capture your interest and motivate deeper exploration of curious curves.

Chapter 1 contains examples of curves that display intriguing properties. We visit self-similar curves, simple curves, space filling curves, self-intersecting curves, curves with zero area, simple curves with positive area, and generalized curves. Examples display that the attractor of an iterated function system may also be a curve. A generalized Cantor set C_h is constructed with a discussion of its length and the area of $C_h \times C_h$. This construction is used as a reference throughout the text where the set C_h appears in various settings.

In Chapter 2 we consider the Koch curve as a path in the plane by constructing the continuous function which defines it. We find the length of the curve and show that it has a tangent nowhere. We conclude with a discussion of the Cantor function.

In Chapter 3 we revisit some of the examples of Chapter 1 and introduce many other curves. We show that every compact set in the plane (including every curve in the plane) is a continuous image of the Cantor set. Homeomorphisms defined on Cantor sets on the unit interval are used to construct a simple curve in the unit square that has positive area at each point!

In Chapter 4 we explore two examples of generalized Koch curves, again looking at self-intersection properties. A new class of Cantor sets is used to describe the double points in one of the examples.

Chapter 5 develops enough theory of metrics to prove that the Hausdorff metric is a complete metric on the set of compact sets in the plane.

Chapter 6 introduces contraction maps and a class of contraction maps called iterated function systems (IFS). Some fractal images are shown to be the fixed points (attractors) of an IFS. We prove that a connected attractor of an IFS is a curve. We also give a necessary and sufficient condition for an attractor to be connected.

Chapter 7 contains an introduction to Hausdorff dimension. Examples of sets of various dimensions are provided by Cantor sets and curves. Tent maps are also discussed. Resident sets for tent maps are either the interval $[0, 1]$ or Cantor sets that generate simple curves with area equal to zero and specified Hausdorff dimension; tangent lines to these curves are also addressed.

Chapter 8 discusses resident sets for quadratic maps. It contains a brief look at fixed points and repellers, Julia sets, and the Mandelbrot set. A Julia set is either connected or it is a Cantor set. We conclude with a question about connected Julia sets.

The material in the appendices is a reference for mathematical properties used in this text that may be unfamiliar to you.

Appendix A contains properties of points on a line and material about sequences and convergence. A few needed details are given about representation of numbers in differing bases.

Appendix B, an intuitive discussion of length and area, provides enough information to be able to find lengths of curves and areas of some interesting and unusual sets in the plane.

In Appendix C maps and sequences of maps are defined and the properties of continuity and convergence discussed. Pertinent topological properties of sets in the plane are discussed.

Appendix D introduces infinite sets and explains why the rational numbers are countably infinite and the real numbers are uncountably infinite.

This book is suitable for a topics course, capstone course, or senior seminar; it is also intended for independent study by students and others interested in mathematics. The topics of this book have proved to be a rich source of projects for undergraduate research for students in a mathematics Research Experiences for Undergraduates (REU) program at the University of Akron. Curves can often provide a better representation of natural phenomena than do the figures of classical geometry. Thus the material is appropriate not only for people working in mathematics, but also those in other sciences. Problems play a vital role in the book. Some are routine, others are more challenging. Occasionally, easily established results used in the text have been made into problems. At other times, proofs of topics not covered in the text are sketched and you are asked to fill in details. Much of the learning of this material will be gained by working through the problems. Solutions to selected problems are provided at the end of the text.

Most notation used is either explained in the text, or else taken from calculus and set theory. A few reminders and additional explanations are collected here.

(1) Natural numbers: $\mathbb{N} = \{1, 2, 3, \cdots\}$.
(2) Integers: $\mathbb{Z} = \{\cdots, -3, -2, -1, 0, 1, 2, 3, \cdots\}$.
(3) Real numbers: $\mathbb{R} = (-\infty, \infty)$.
(4) The plane: $\mathbb{R}^2 = \mathbb{R} \times \mathbb{R} = \{(x, y) : x, y \in \mathbb{R}\}$.
(5) Complex numbers: $\mathbb{C} = \mathbb{R}^2$ also considered as points in the plane.
(6) Intervals in $\mathbb{R}$: $(a, b) = \{x : a < x < b\}$ and $[a, b] = \{x : a \leq x \leq b\}$.
(7) Intervals in $\mathbb{R}^2$: For points $z, w \in \mathbb{R}^2$ the interval $[z, w]$ is the line segment connecting z and w with parametric representation $(1 - t)\, z + tw$, $t \in [0, 1]$.

We will use the notion of a function, the definition of a continuous function, the sum of an infinite series, the limit of a sequence, the least upper bound property for the real number system, uniformly Cauchy sequences of functions, and a bit about countable and uncountable sets. The formal prerequisite is a strong course in calculus with some experience at reading and writing proofs. Elementary analysis is extremely efficient for investigating many features of curves.

Richard B. Darst, Judith A. Palagallo, and Thomas E. Price

Acknowledgments

Teachers can learn from their students. We are happy to have comments from some who watched this material evolve over the past few years. Thanks are due to the University of Akron REU students who sat through many lectures on these topics. Michael Cantrell earns special thanks for reading the entire manuscript of an early version. The kind hospitality of Isla Kercher and Timberline Church in Fort Collins provided a friendly and comfortable environment where our writing team could work. In addition, the Science and Fractal Analysis Research Institute at Crystal Lakes Colorado provided an excellent location for contemplation. Darst gladly dedicates his contribution to this work to Jeane. Palagallo and Price thank their family members who patiently watched this project from its inception. We thank Jianping Zhu for his inspiration and continuing encouragement. Ed Dunne of the American Mathematical Society worked with us over several months to reorganize and clarify the material. The book is much better because of his suggestions. We thank Dorian Mueller, Rajesh Babu, and Yee Sern Tan of World Scientific for taking on this project, their editorial skill, and helpful advice. Dr. D. P. Story of AcroTEXprovided valuable LATEXstyles and advice.

Contents

List of Figures

List of Tables

Chapter 1

Examples of curious curves

While curves in the plane have intrigued mathematicians for centuries, the modern era erupted from Georg Cantor's investigations that began in the 1860's. Cantor showed that a line and a plane contain the same number of points by displaying a map from the *unit interval* I onto the *unit square* $U = I \times I$. Following this discovery, the natural question "Can you find a continuous map of I onto U?" drew the interest of many prominent mathematicians. Reference [Dauben (1970)] contains a fascinating discussion of Cantor's work, including, of course, the classical middle-thirds Cantor set that appeared in print as a note in 1883.

In this book a *curve* is the image of a nontrivial continuous map from I into the plane. (The image of a constant map is one point, a trivial curve.) The first example of a curve that contains a square, a *space filling curve*, appeared in 1890. Numerous interesting examples of space filling curves appear in [Sagan (1994)] with displays, photos, and commentary. A *simple curve* is the image of a one-to-one, continuous map from I into the plane.

Our central focus is the question "What can we learn about curves?" Elementary analysis is a basic tool for addressing this question. For example basic analysis results show that a curve is a compact, connected set in the plane and that a simple curve can contain no square. So U is a curve, but not a simple curve. Simple curves can be quite complex, possessing surprising properties related to area, Hausdorff dimension (which lets us compare sizes of curves that have area equal to zero), and smoothness. Geometry and Cantor sets are also basic tools for studying the subtleties of such curves.

Felix Klein said "Everyone knows what a curve is until he has studied enough mathematics to become confused through the countless number of possible exceptions." We aim to present a sampling of the classical results and introduce you to some fascinating new sets, some of which are not curves but are related. In most cases we illustrate the construction of the curves with computer generated drawings. In this chapter we begin with examples that illustrate particular features of curves and then introduce a class of Cantor sets that we use frequently. The interplay between curious curves and Cantor sets is a central theme of this book.

1.1 Variations of the Koch curve

The Koch curve was introduced by the Swedish mathematician Helge von Koch in 1904 as an example of a curve that does not have a tangent line at any point [Edgar (1993)]. It is also a classical example of a *self-similar curve*, in which every piece contains a portion similar to the whole figure.

1.1.1 *Koch curve*

The Koch curve is a self-similar, simple curve. There are several ways to describe the Koch curve. In this introduction, we construct the curve by an iterative replacement process beginning with the generator shown in Figure 1.1. We replace each of the four linear segments (parts) of the generator with a $1/3$-scaled copy of the generator as shown in Figure 1.2. When the process is repeated ad infinitum, the limiting image is defined to be the Koch curve K. A representation of the curve is shown in Figure 1.3. In Chapter 2 we will give the details of this construction and will define the function that gives K as a one-to-one continuous image of I. Note that the Koch curve is composed of four similar images of itself. Each similar image lies along a side of the generator. The Koch curve does not have a tangent at any point. Intuitively, the Koch curve has area equal to zero. We will later see simple curves that have non-zero area, which seems counter-intuitive (at this point).

The Koch snowflake (See Figure 1.4.) is obtained by replacing each of the three edges of an equilateral triangle with an outward facing Koch curve. The Koch snowflake is neither a self-similar curve nor a simple curve; it is a one-to-one, continuous image of a circle.

1.1.2 *Modified Koch curve*

Generalizations of the Koch curve begin with a generic generator $G = G(a, \theta)$ as shown in Figure 1.5, allowing the angle θ and the length a to vary. Note that the length L in this figure depends on a and θ. If $L < 1$ then G generates a self-similar curve by a replacement process similar to that for the Koch curve described above. In Figure 1.6 we show modifications for a few values of a with θ fixed at $\pi/3$.

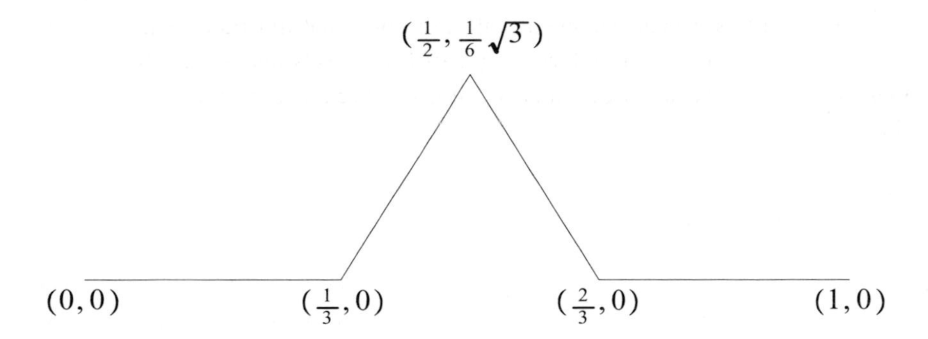

Fig. 1.1: Graph of the generator of the Koch curve.

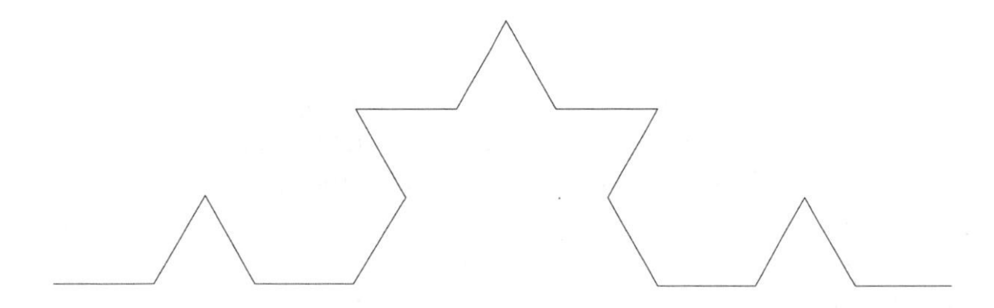

Fig. 1.2: Stage 2 in the generation of the Koch curve.

Note especially that for $a > 1/4$ the curve appears to be a simple curve. However, for values of $a < 1/4$, we see that the curve is self-intersecting. Because of this property, we designate $a = 1/4$ as the *pivotal value*. The curve generated with $a = 1/4$ displays the just-touching property. The self-intersection points of this curve are the vertices of equilateral triangles that are replicated throughout the curve because of the self-similarity property.

With the value of $\theta = \pi/4$, there is also a unique pivotal value which allows for the self-intersecting curve shown in Figure 1.7. We will show that the points of self-intersection in this curve form Cantor sets of points in the plane. These modified curves maintain many of the properties of the classic Koch curve while displaying intriguing differences. The details to verify that the modified Koch curve is a continuous image of I are given in Section 4.1.

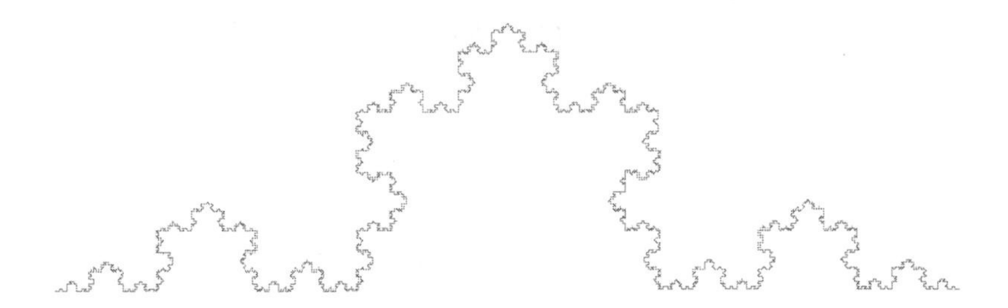

Fig. 1.3: The Koch curve.

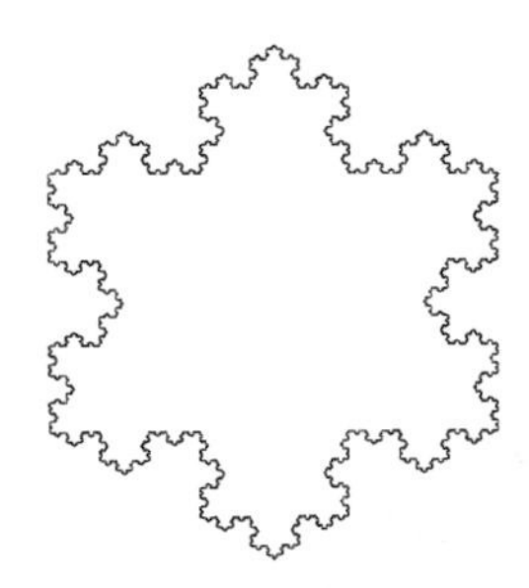

Fig. 1.4: The Koch snowflake.

1.1.3 *Basics of complex numbers*

As the Koch curve indicates, we extensively use rotations in the plane, conveniently described by multiplication of complex numbers. The set $\mathbb{C} = \{x + iy : x, y \in \mathbb{R}\}$ of complex numbers is closely related to the Euclidean plane

$$\mathbb{R}^2 = \{(x, y) : x, y \in \mathbb{R}\}.$$

We use the identification $x + iy \Leftrightarrow (x, y)$ and go between $\mathbb{R}^2$ and $\mathbb{C}$ without comment. Let $z = x+iy$ and $w = u+iv$ be in $\mathbb{C}$. Then addition and multiplication are defined on $\mathbb{C}$ by

$$z + w = (x + u) + i\,(y + v)$$

and

$$zw = (x + iy)\,(u + iv) = (xu - yv) + i\,(xv + yu) \tag{1.1}$$

respectively. In particular, $i^2 = -1$. Notice that addition and multiplication of real numbers are preserved. The modulus of z is the real number

$$|z| = |x + iy| = |(x, y)| = \sqrt{x^2 + y^2}.$$

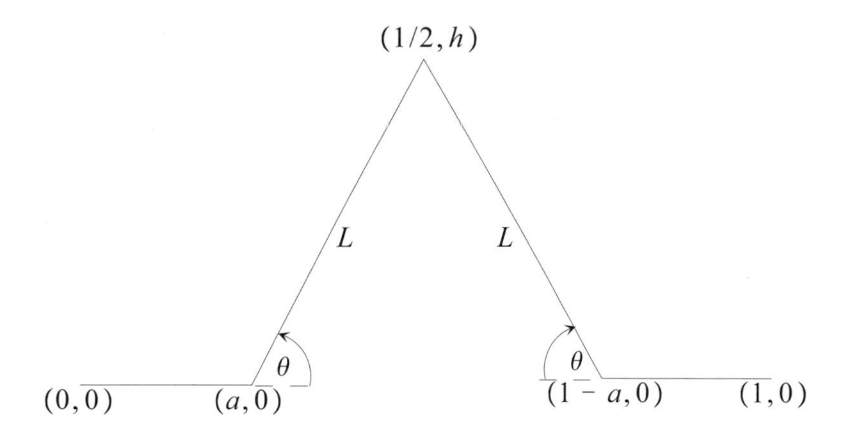

Fig. 1.5: Generic generator.

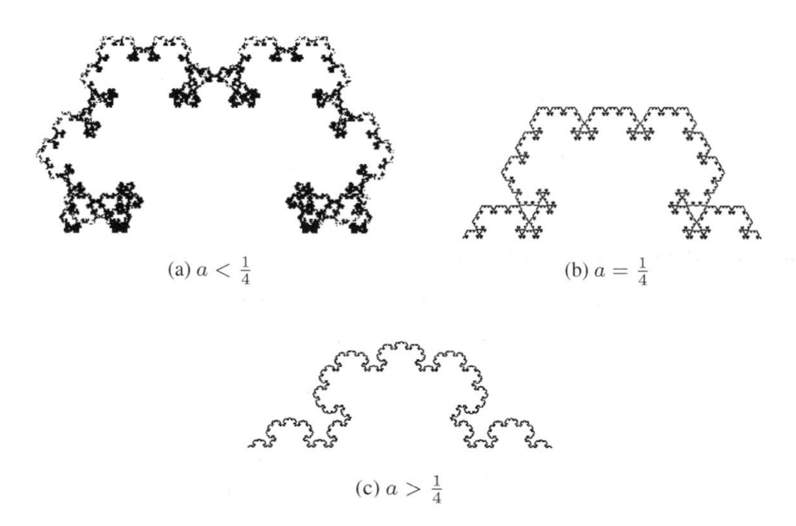

(a) $a < \frac{1}{4}$

(b) $a = \frac{1}{4}$

(c) $a > \frac{1}{4}$

Fig. 1.6: Generalized Koch curves.

Given $z \in \mathbb{C}$, it follows from trigonometry that there are numbers $\theta \in \mathbb{R}$ satisfying the equations $x = |z| \cos\theta$ and $y = |z| \sin\theta$. If $z = 0$, the set of solutions is $\mathbb{R}$. Otherwise, there is one solution, say ϕ, in $[0, 2\pi) = \{t \in \mathbb{R} : 0 \leq t < 2\pi\}$ and the other solutions are of the form $\theta = \phi + 2j\pi$, where j is any integer; the solutions comprise an equivalence class $\{\theta : \theta = \phi + 2j\pi,\ j \in \mathbb{Z}\}$.

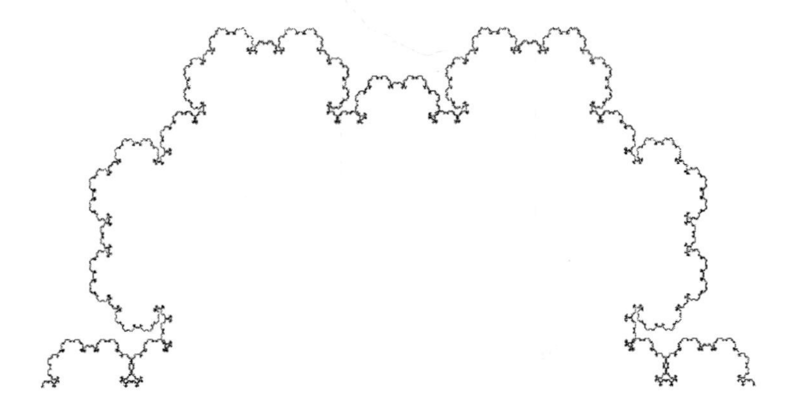

Fig. 1.7: Example of a self-intersecting curve with $\theta = \pi/4$.

We will usually suppress the equivalence class notation and simply let any number in the class stand for the whole class.

A complex number $(\cos\theta, \sin\theta)$ of modulus 1 is given the exponential notation $e^{i\theta}$. Complex exponentials behave like real exponentials. For example, if $z = x+iy = |z|(\cos\theta+i\sin\theta) = |z|e^{i\theta}$ and $w = u+iv = |w|(\cos\phi+i\sin\phi) = |w|e^{i\phi}$, using Equation 1.1 and the addition formulas for sine and cosine, we have

$$\begin{aligned}
zw &= |z|\,e^{i\theta}\,|w|\,e^{i\phi} \\
&= |z|\,|w|\,e^{i\theta}e^{i\phi} \\
&= |z|\,|w|\,(\cos\theta + i\sin\theta)(\cos\phi + i\sin\phi) \\
&= |z|\,|w|\,((\cos\theta\cos\phi - \sin\theta\sin\phi) + i\,(\sin\theta\cos\phi + \cos\theta\sin\phi)) \\
&= |z|\,|w|\,(\cos(\theta+\phi) + i\sin(\theta+\phi)) \\
&= |z|\,|w|\,e^{i(\theta+\phi)}.
\end{aligned}$$

In particular, multiplying z by $e^{i\phi}$ rotates the point z by an angle ϕ about the origin in a counter-clockwise direction when $\phi > 0$.

1.2 More examples of curious curves

1.2.1 *The unit square is a curve*

We define an initiator $f_0(t) = (t,t)$, the line segment in $U = [0,1] \times [0,1]$ connecting the points $(0,0)$ and $(1,1)$. The first replacement step $f_1(I)$, tracing the directed path from $(0,0)$ to $(1,1)$ following the numerical order of the squares

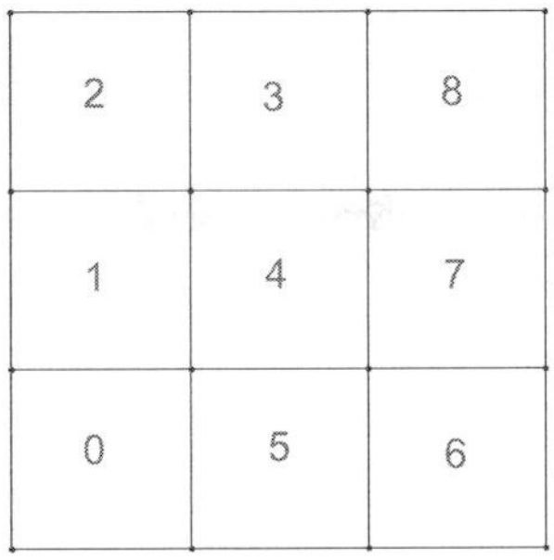

2	3	8
1	4	7
0	5	6

Fig. 1.8: Strategy for defining f_1.

in Figure 1.8, is shown on the left of Figure 1.9. If the replacement scheme is continued with rotations as suggested by $f_2(I)$ on the right of Figure 1.9, the limiting curve will fill the entire square. Other illustrative drawings and the proof that every element in U is the image of a point in I are deferred to Chapter 3. Note also that U is a curve with area one.

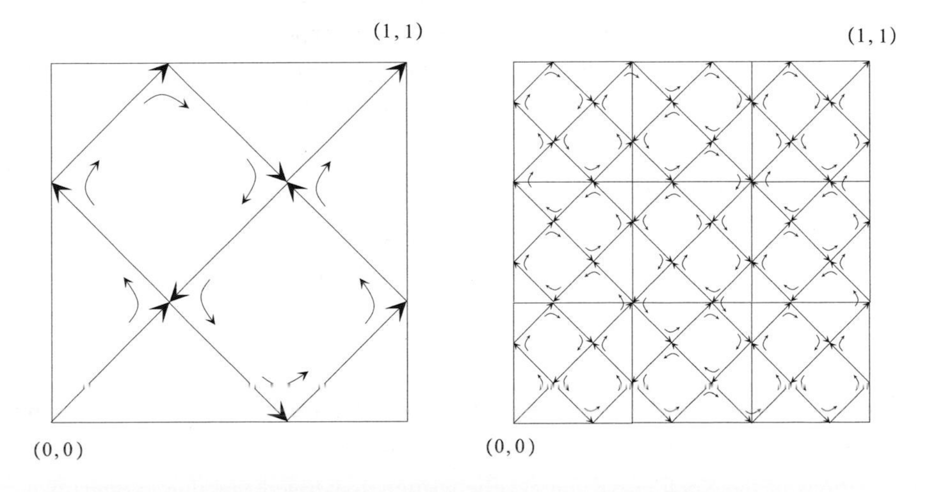

Fig. 1.9: Continuously mapping the unit interval onto the unit square.

1.2.2 *Iterated function systems produce curves*

The word *fractal*, coined by Mandelbrot [Mandelbrot (1983)], appears throughout our discussion of curves. Many fractals have the self-similarity property; that is, they are made up of parts which are, in some way, similar to the whole. For example, the Koch curve is made up of four scaled copies of itself. Fractal images can be generated by replacement strategies. A more complete discussion of this material is in Chapter 6, but we introduce some fundamental ideas here.

Let D be a closed subset of $\mathbb{R}^2$. A function f defined on D to $\mathbb{R}^2$ is called a **contraction map** on D if there exists a real number $e < 1$ such that $|f(s) - f(t)| \leq e\,|s - t|$ for each pair (s, t) in D.

Definition 1.1. An **iterated function system** (IFS) is a collection of contraction mappings $\{f_i,\ i = 1, \ldots, m\}$. A compact set F is **invariant** for the transformations $\{f_i\}$ if

$$F = \cup_{i=1}^{m} f_i\,(F)\,.$$

Invariant sets are often fractals.

In Chapter 6 we will see that an IFS has exactly one invariant set.

Example 1.1. The Koch curve K can be described as an invariant set. Let $f_1, f_2, f_3, f_4 : I \rightarrow \mathbb{R}^2$ be the four functions

$$f_1\,(z) = (1/3)\,z,\ f_2\,(z) = (1/3) + (1/3)\,e^{i\pi/3}z$$
$$f_3\,(z) = \left(1/2, \sqrt{3}/6\right) + (1/3)\,e^{-i\pi/3}z,\ f_4\,(z) = (2/3) + (1/3)\,z.$$

Then $f_1\,(I)\,, f_2\,(I)\,, f_3\,(I)$ and $f_4\,(I)$ are the four pieces of the generator in Figure 1.1. Furthermore, $f_1\,(K)\,, f_2\,(K)\,, f_3\,(K)$ and $f_4\,(K)$ are the four similar pieces of K so that

$$K = f_1\,(K) \cup f_2\,(K) \cup f_3\,(K) \cup f_4\,(K)\,.$$

Thus K is invariant for the mappings of the IFS $\{f_1, f_2, f_3, f_4\}$.

In Chapter 2 we approach the Koch curve from a different perspective. There we generate a nested sequence $\{E_k\}$ of compact sets that converge to K.

The previous example demonstrates that an IFS can produce a curve. Modifications of the Koch curve can also be written as iterated function systems that produce curves. Other examples of this process will be discussed in detail in Chapter 6 where we define the continuous function whose image is the invariant set. There we will develop the necessary definitions and topological and convergence properties. The culminating theorem for that discussion is

Theorem 1.1. *A connected invariant set (attractor) of an iterated function system is a curve.*

Example 1.2. The Sierpinski triangle is a curve. The Sierpinski triangle is the invariant set for three transformations in the plane. These three transformations define the process of replacing the equilateral triangle shown in Figure 1.10a with the three scaled equilateral triangles shown in Figure 1.10b. Repeating the replacement process produces an image represented in Figure 1.10c. By the theorem, the Sierpinski triangle is a curve. It is connected because each iterate is connected. (See Theorem 6.6.)

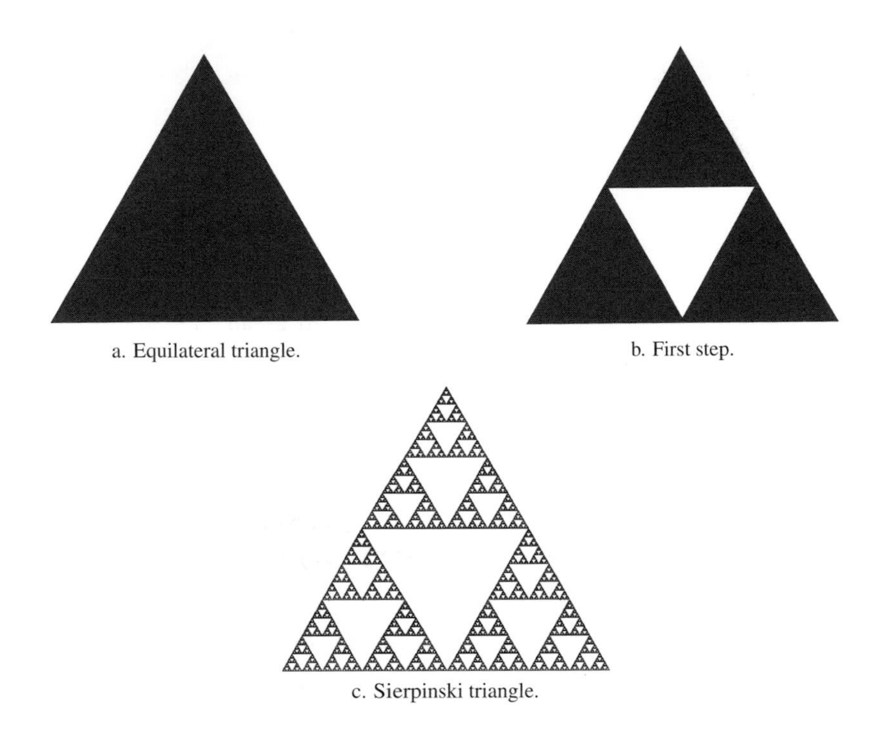

a. Equilateral triangle.

b. First step.

c. Sierpinski triangle.

Fig. 1.10: Construction of the Sierpinski triangle.

Example 1.3. We can modify the construction in the previous example and replace the original equilateral triangle with the image in Figure 1.11a. The result, another curve, is the irregular Sierpinski triangle represented in Figure 1.11b.

Example 1.4. Begin with the unit square and define 16 functions that will replace the square with the sixteen smaller squares shown in Figure 1.12a. Repeating the replacement process will produce the image shown in Figure 1.12b, another

a. Original triangle.

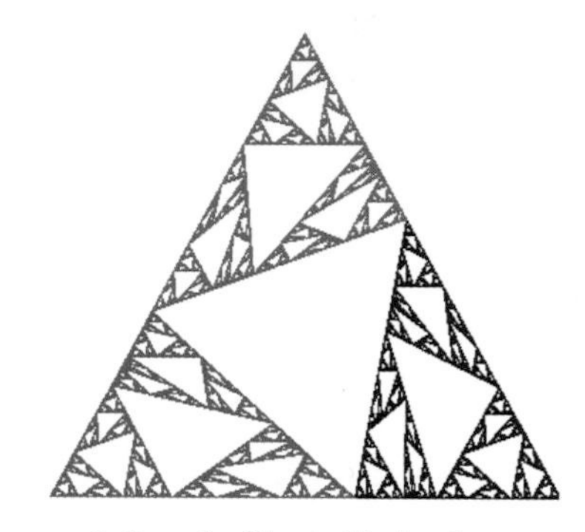

b. Irregular Sierpinski triangle.

Fig. 1.11: Construction of the irregular Sierpinski triangle.

example of a curve. An exercise in Chapter 6 asks you to find the functions needed to produce the drawing. The limiting figure is called the "Badge and Hydrant" or "hydrant" curve.

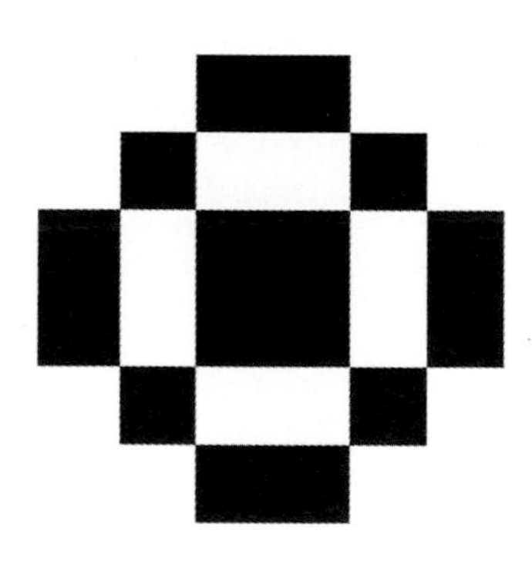

a. First iteration.

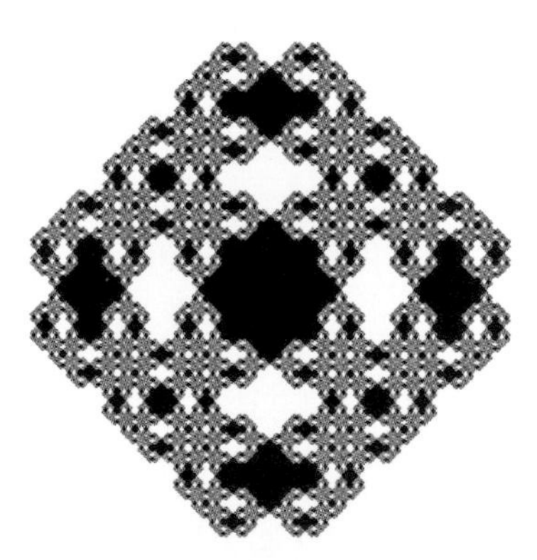

b. Badge and Hydrant curve.

Fig. 1.12: Construction of the Badge and Hydrant curve.

1.3 Construction of a family of Cantor sets

In this section we construct a family of Cantor sets. When such sets first appeared in the mathematical literature, they were categorized by some mathematicians as pathological and only of interest to those who studied mathematical oddities. They are now known to be appropriate tools for analyzing the behavior and structure of sets.

1.3.1 *Middle-thirds Cantor set*

Before constructing general Cantor sets, we construct the classical middle-thirds Cantor set. It is one of the best known and most easily constructed fractals, and it displays many typical fractal characteristics. We denote the classical middle-thirds Cantor set by C . This subset of $[0, 1]$ is obtained by successive deletion of middle-third open subintervals as follows:

$$\begin{aligned} I_0 &= [0, 1] \\ I_1 &= [0, 1/3] \cup [2/3, 1] \\ I_2 &= [0, 1/9] \cup [2/9, 1/3] \cup [2/3, 7/9] \cup [8/9, 1] \\ &\vdots \end{aligned}$$

Note that I_n is I_{n-1} with the middle open third of each subinterval removed. For each $n \in \mathbb{N}$, I_n is the union of 2^n closed intervals, each of length $1/3^n$, and $I_0 \supset I_1 \supset I_2 \supset \cdots$. We define $C = \cap_{n\geq 0} I_n$. The construction is illustrated in Figure 1.13.

The Cantor set C can also be described as the invariant set for the IFS $\{f_1, f_2\}$, where

$$f_1(x) = \frac{1}{3}x;\ f_2(x) = \frac{1}{3}x + \frac{2}{3}.$$

Then $f_1(C)$ and $f_2(C)$ are just the left and right 'halves' of C, so that $C = f_1(C) \cup f_2(C)$. Thus C is invariant for the mappings f_1 and f_2. In Chapter 3 we will show the remarkable fact that any compact set in $\mathbb{R}^2$ is a continuous image of C.

Because we begin the construction of C with three subintervals of I, it is convenient to use base 3 representation of elements in I. Any number $x \in I$ has a base 3 expansion of the form $x = \sum_{j=1}^{\infty} x_j/3^j$, where $x_j \in \{0, 1, 2\}$; $x = 0.x_1x_2\cdots$ base 3. For example, $1/3 = 0.1$ base 3. Normally, we use an underbar to denote continuing repetition of digits in a base 3 expansion; and so, using the properties of convergent geometric series, we see that $1/3 = 0.0\underline{2}$ base 3 also. Any base 3 representation of a number for which $x_j = 0$ for all j larger than some positive integer N is called a terminating expansion. A number with a terminating expansion has two base 3 representations. For example, $2/3 = 0.2\underline{0}$ base 3 $= 0.1\underline{2}$ base 3. In such cases you can use whichever expansion is convenient. With this exception it can be shown that every $x \in I$ has a unique base 3 representation. From the construction of C, we see that each element of C can be represented as a string of 0's and 2's. Furthermore, every string of 0's and 2's represents a point in C. For example $1/4 = 0.\underline{02}$ is an element of C.

Fig. 1.13: Initial steps in the construction of the Cantor set.

1.3.2 *Construction of generalized Cantor sets*

You will frequently be encouraged to refer to Figure 1.14 where we have labeled the x-coordinates of a few points that are referenced in the construction. You should be able to determine any unlabeled points. To construct the sets depicted in Figure 1.14 we begin with the unit square

$$U = [0, 1] \times [0, 1] = \{(x, y) : 0 \leq x \leq 1, 0 \leq y \leq 1\}$$

in the plane and remove a sequence of triangles from this square according to the following steps.

Example 1.5. Construction of a family of Cantor sets.

Step 1 Remove the inside of the triangle in Figure 1.14a with vertices $\left(\frac{1}{3}, 0\right), \left(\frac{1}{2}, 1\right)$ and $\left(\frac{2}{3}, 0\right)$ along with the open interval $(1/3, 2/3)$. Notice that we keep the two line segments $\left[\left(\frac{1}{3}, 0\right), \left(\frac{1}{2}, 1\right)\right]$ and $\left[\left(\frac{2}{3}, 0\right), \left(\frac{1}{2}, 1\right)\right]$ that compose the sides of the triangle and that nothing is removed from the interval $[(0, 1), (1, 1)]$ composing the top edge of U. In what follows we will simply write "Remove the triangle" in situations where we remove the interior and base leaving the base vertices of a triangle. Also, note that removing the interval $(1/3, 2/3)$ from the interval $[0, 1]$ is the first step in the construction of the classical Cantor set. Let S_1, the part of Figure 1.14a shaded light gray, denote the part of the square U that remains after the triangle is removed.

Step 2 Remove 2 triangles with top vertices $\left(\frac{1}{4}, 1\right)$ and $\left(\frac{3}{4}, 1\right)$ and base segments $(1/9, 2/9)$ and $(7/9, 8/9)$, respectively. The two base segments are the open middle thirds of the two base intervals $\left[0, \frac{1}{3}\right]$ and $\left[\frac{2}{3}, 1\right]$ that remain after *Step 1*. Let S_2, shaded light gray in Figure 1.14b, denote the subset of S_1 that remains after these two triangles are removed. Again, nothing is removed from the interval $[(0,1),(1,1)]$.

Step 3 Remove the 2^2 triangles with top vertices $\left(\frac{2j-1}{2^3},1\right)$, $j=1,\ldots,2^2$, and corresponding base segments that are the open middle thirds of the four base intervals of length $\frac{1}{3^2}$ that remain after *Step 2*. Let S_3, shaded gray in Figure 1.14c, denote the subset of S_2 that remains after these 2^2 triangles are removed.

Step n Continue iteratively so that at *Step n* we remove 2^{n-1} triangles, each with a base segment of length $\frac{1}{3^n}$ (an open middle third of a base interval of length $\frac{1}{3^{n-1}}$ that remains after *Step n–1*) and a top vertex $\left(\frac{2j-1}{2^n},1\right)$, $j=1,\ldots,2^{n-1}$, in the middle of a corresponding interval of length $\frac{1}{2^{n-1}}$ on the top. A set S_n remains. *Notice that the points at the vertices remain.*

After completing a step for every positive integer n, we have removed an infinite collection of triangles from U. Denote the set that remains by S. For $0\leq h\leq 1$, denote by C_h the intersection of S with the horizontal line $y=h$, which is distance h above the base of the square. (The line $y=h$ is the set $\{(x,h):x\in\mathbb{R}\}$.) Figure 1.14c shows the line for $h=1/3$.

(1) When $h=0$, we have the part of the base of U which has not been removed; this set C_0 is the classical (standard, middle-thirds) Cantor set in the interval $[0,1]$.
(2) When $0<h<1$, C_h is a Cantor set on the interval $[(0,h),(1,h)]=\{(x,h):x\in I\}$.
(3) When $h=1$, we have the entire top edge of U.

1.3.3 *The length of the Cantor set C_0*

We begin by setting $h=0$ so that we will calculate the length of the standard middle-thirds Cantor set. At **Step 1**, a single segment of length $1/3$ is removed from the interval $[0,1]$ on the x-axis. At **Step 2** two segments of length $1/3^2$ are removed so that the total length of the removed segments is $1/3+2(1/3^2)=5/9$. In general, at **Step n**, 2^{n-1} segments of length $1/3^n$ are removed. The total length ℓ_0 of the segments removed is

$$\ell_0=\frac{1}{3}+2\frac{1}{3^2}+2^2\frac{1}{3^3}+\cdots=\frac{1}{3}\left(1+\frac{2}{3}+\left(\frac{2}{3}\right)^2+\cdots\right)=1.$$

Thus, the length of the Cantor set is

$$\operatorname{len}(C_0)=\operatorname{len}([0,1])-\ell_0=0.$$

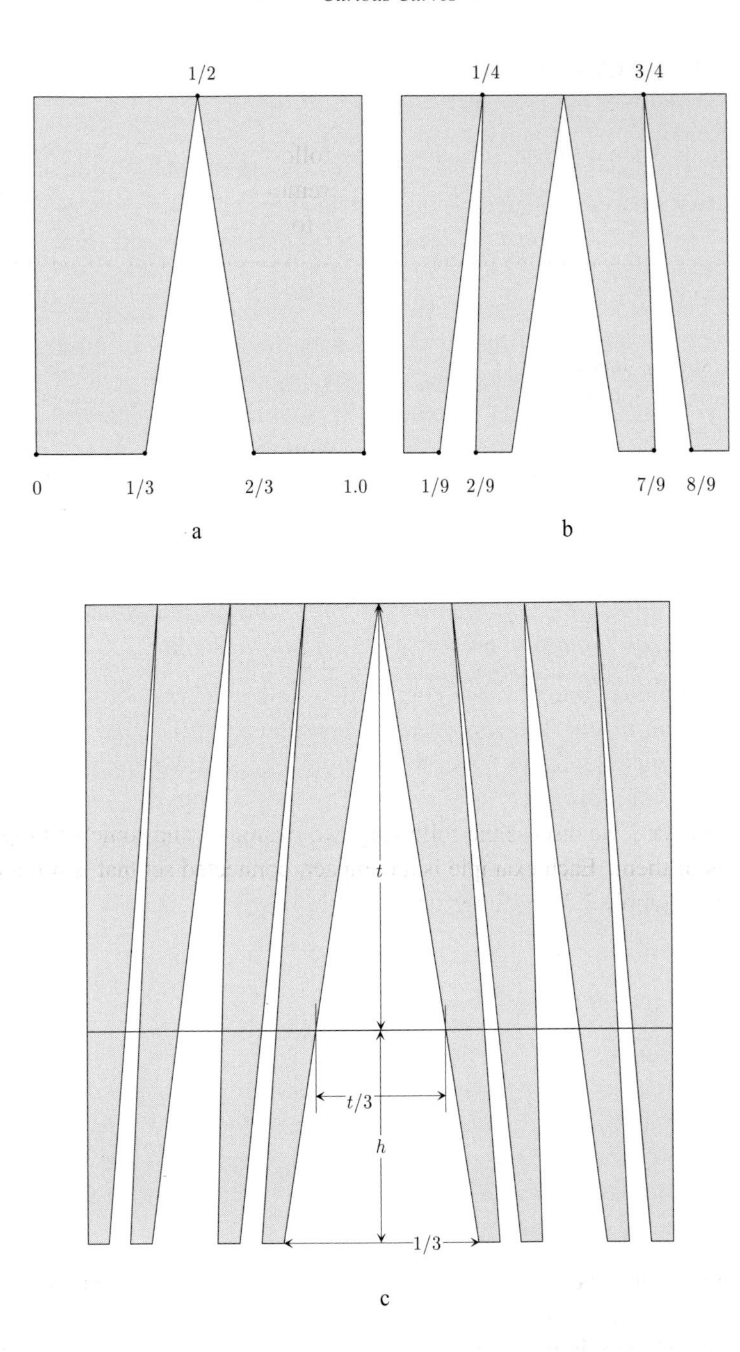

Fig. 1.14: A family of Cantor sets.

1.3.4 *The sets C_h*

Now put $t = 1 - h$ and examine Figure 1.14c which, coupled with the above calculations, should provide insights to the following discussion. At **Step n**, the length of each of the 2^{n-1} segments that are removed from the line $y = h$ is $t/3^n$. Thus (referring to the calculation given above for the case $t = 1$ or $h = 0$), t is the total length of the segments that are removed from the interval $[(0,h),(1,h)] \subset U$. Consequently, $\text{len}(C_h) = 1 - \ell_t = 1 - t = h$. In particular, $\text{len}(C_1) = 1$ which is not surprising since $C_1 = I$.

The set C_h has length h and contains the same number of points as I, but it contains no interval when $0 \leq h < 1$. We will use these properties of the sets C_h to construct curves in the plane. In Chapter 3 we further verify that $\text{area}(C_h \times C_h) = h^2$ and use this set product to construct simple curves with positive area.

1.4 What is not a curve?

The standard Cantor set $C = C_0$ shares several properties with the interval $[0,1]$. Both sets are compact, and they contain the same number of points. However, they differ drastically with respect to connectedness; while I is connected, the only connected subsets of C are sets containing a single point. By Definition C.17, C is totally disconnected. The Cantor set is not a curve.

In Chapter 3 we discuss the following two examples and some striking modifications of them. Each example is a compact, connected set that is not a curve. The first (Example 3.3) is the geometric comb

$$GC = I \cup \{(0,y) : 0 \leq y \leq 1\} \cup \bigcup_{n \geq 0} \{(1/2^n, y) : 0 \leq y \leq 1\}.$$

(See Figure 1.15.)

A *generalized curve* is a continuous image of the closed half line $[0,\infty)$. We will demonstrate that GC is a generalized curve.

The second (Example 3.4) is the Cantor comb

$$CC = I \cup \bigcup_{x \in C} \{(x,y) : 0 \leq y \leq 1\} = I \cup (C \times I),$$

where the bases of the tines are the points in the Cantor set. We will show that the Cantor comb is not a generalized curve but as a seemingly minor modification of CC, shown in Figure 1.16, is a curve. Likewise, by the end of Chapter 3 you will be able to define a continuous map from I onto the curve illustrated by Figure 1.17. This curve is composed of the unit circle R_1 and a sequence of larger

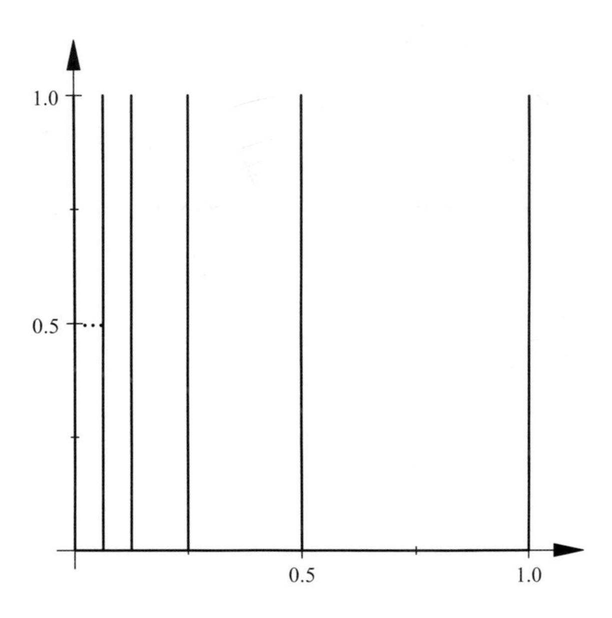

Fig. 1.15: The geometric comb is an example of a compact, connected set that is not a curve.

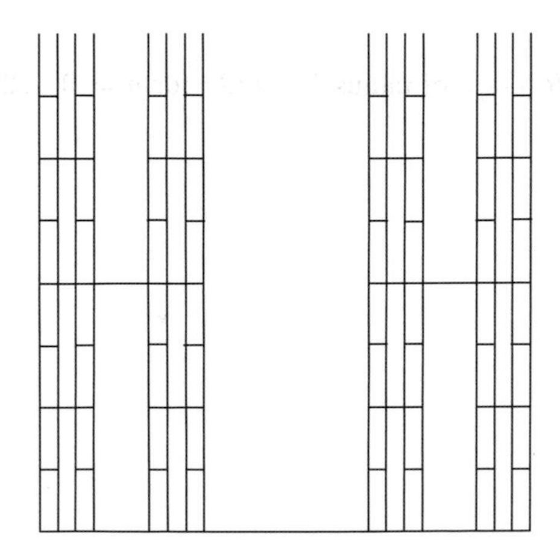

Fig. 1.16: Modified Cantor comb.

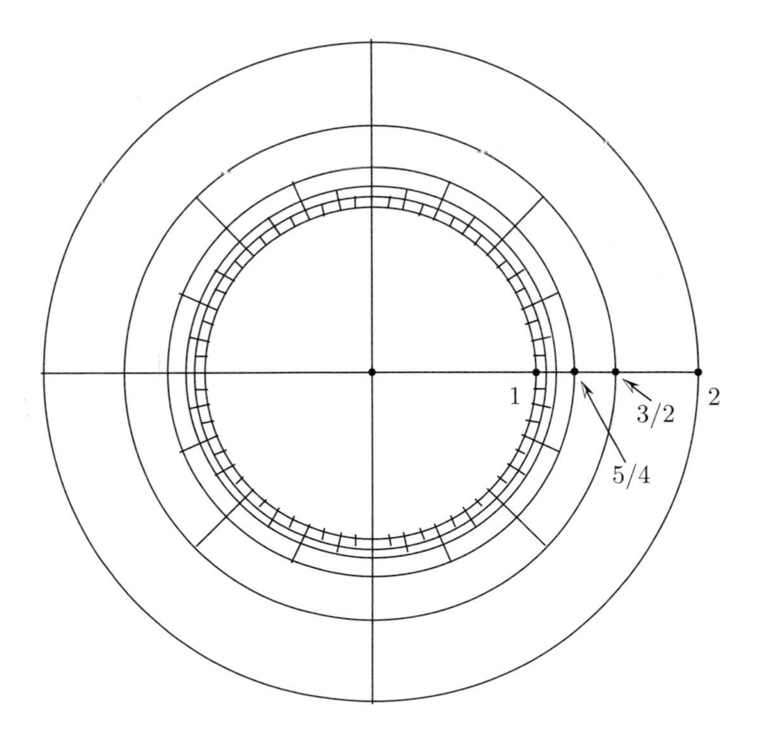

Fig. 1.17: A circular shaped curve.

concentric circles $R_{1+1/2^n}$ of radius $1 + 1/2^n$ for $n = 0, 1, 2, 3, 4, \cdots$, together with radial spokes connecting the unit circle to the larger circles. The first six circles and their spokes are displayed in Figure 1.17.

Chapter 2

The Koch curve and tangent lines

The Koch curve shown in Figure 2.1 is a classical example of a self-similar curve. The rich history of such curves dates at least to 1904 when Helge von Koch first described the curve that bears his name. The curve was designed by him to be an example of a curve with no tangent lines. In this chapter we explore properties of the Koch curve and give a functional expression that describes it. We also define tangent lines to simple curves in general and then show that the Koch curve does not have a tangent line at any point. Next, we present an example of a continuous function from I to I whose graph has no tangent lines. Finally, we construct the Cantor function, a continuous map from from I to I that maps the Cantor set continuously onto I; its graph has a vertical tangent line at many points.

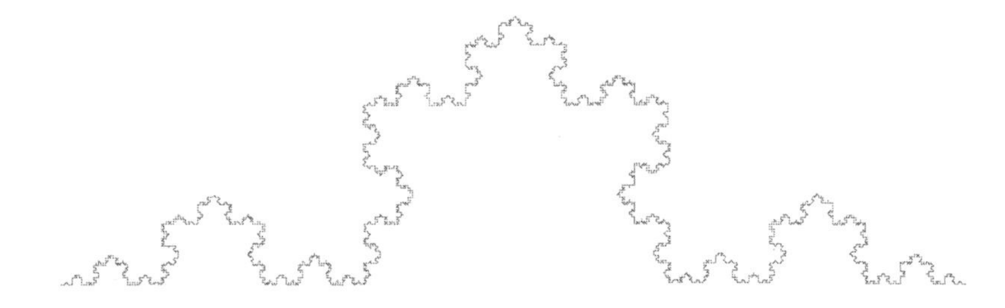

Fig. 2.1: Koch curve.

2.1 Construction of the Koch curve

In this section we show that the Koch curve introduced in Subsection 1.1.1 of Chapter 1 is the image of a continuous function $f : I \longrightarrow \mathbb{R}^2$. The function f is developed by constructing a uniformly Cauchy sequence of continuous functions $f_k : I \longrightarrow \mathbb{R}^2$ that converges to f. Thus, f is continuous and $f(I)$ is a *curve* in $\mathbb{R}^2$.

We begin with an **initiator** $f_0(x) = x = (x, 0)$ and recursively define the sequence $\{f_k\}$ of continuous maps from I to $\mathbb{R}^2$. The image of f_1 (Figure 2.2) is called the **generator**.

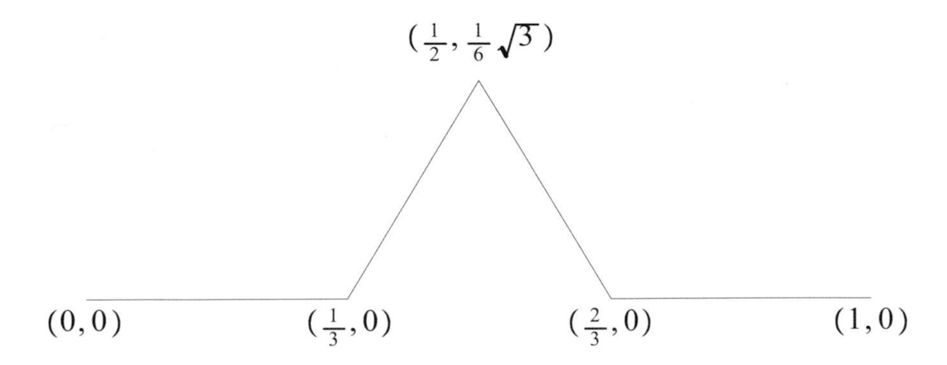

Fig. 2.2: Graph of $f_1(I)$, the generator of the Koch curve.

2.1.1 *Representation in* base 4

Because the generator is composed of 4 intervals, we use base 4 representations of numbers in I. Any number $x \in I$ has a base 4 expansion of the form

$$x = 0.x_1 x_2 x_3 \cdots \text{base } 4,$$

where $x_j \in \{0, 1, 2, 3\}$ and $x = \sum_{j=1}^{\infty} x_j / 4^j$. For example, using the properties of a convergent geometric series we have $1/3 = 0.111 \cdots \text{base } 4$ since $1/3 = \sum_{j=1}^{\infty} 1/4^j$. Consequently, $2/3 = 0.\underline{2}$ where we use an underbar to denote continuing repetition of digits in a base 4 expansion. Using this notation we have

$$\begin{aligned} 0.1\underline{2} \text{ base } 4 &= 0.1 \text{ (base } 4) + 0.0\underline{2} \text{ (base } 4) \\ &= (1/4) + (1/4)(0.\underline{2} \text{ (base } 4)) \\ &= (1/4) + (1/4)(2/3) = 5/12. \end{aligned}$$

A number with a terminating expansion has two base 4 representations. For example, $1/2 = 0.2\underline{0} \text{ base } 4 = 0.1\underline{3} \text{ base } 4$. In such cases you can use whichever expansion is convenient. With this exception every $x \in [0, 1]$ has a

Table 2.1: Values of $C(j)$ for the Koch curve.

j	0	1	2	3	4
$C(j)$	$(0,0)$	$(1/3,0)$	$(1/2,\sqrt{3}/6)$	$(2/3,0)$	$(1,0)$

unique base 4 representation. For simplicity of notation we will often abbreviate $x = 0.x_1x_2\cdots$ base 4 by omitting the quantifier "base 4" and simply write $x = 0.x_1x_2\cdots$ when there is no confusion.

2.1.2 *Formulas for f_k*

We make use of the left shift operator ℓx defined for $x = 0.x_1x_2x_3\cdots$ base $4 \in I$ by

$$\ell x = 0.x_2x_3\cdots \text{ base } 4.$$

Notice that ℓ maps each of the four intervals $[j/4,(j+1)/4]$, $j = 0,1,2,3$, linearly onto I.

We define a set of 5 points C_j, $j = 0,1,2,3,4$, by the values in Table 2.1. We also define

$$D(j) = C(j+1) - C(j),\ j = 0,1,2,3 \tag{2.1}$$

so

$$D(0) = \frac{1}{3},\ D(1) = \frac{1}{3}e^{\frac{i\pi}{3}},\ D(2) = \frac{1}{3}e^{\frac{-i\pi}{3}},\ \text{and } D(3) = \frac{1}{3}.$$

For $k \geq 1$ put

$$f_k(x) = C(x_1) + D(x_1)f_{k-1}(\ell x),\ k \geq 1. \tag{2.2}$$

Example 2.1. Since $f_1(0.x_1x_2x_3\cdots) = C(x_1) + D(x_1)f_0(0.x_2x_3\cdots)$,

$$\begin{aligned} f_1(5/12) &= f_1(0.1\underline{2}\text{ base } 4) \\ &= C(1) + D(1)f_0(0.\underline{2}\text{ base } 4) \\ &= \frac{1}{3} + \left(\frac{1}{6},\frac{\sqrt{3}}{6}\right)\left(\frac{2}{3}\right) \\ &= \frac{4}{9} + \frac{\sqrt{3}}{9}i. \end{aligned}$$

Example 2.2. Likewise, $f_2(0.x_1x_2x_3\cdots) = C(x_1) + D(x_1)f_1(0.x_2x_3\cdots)$. Hence,

$$\begin{aligned} f_2(29/48) &= f_2(0.21\underline{2}\,\text{base}\,4) \\ &= C(2) + D(2)f_1(.1\underline{2}\,\text{base}\,4) \\ &= \frac{1}{2} + i\frac{\sqrt{3}}{6} + \left(\frac{1}{6} - i\frac{\sqrt{3}}{6}\right)\left(\frac{4}{9} + \frac{\sqrt{3}}{9}i\right) \\ &= \frac{17}{27} + \frac{\sqrt{3}}{9}i. \end{aligned}$$

Examining the images of the maps $f_k, k = 1, 2, \ldots$ provides additional insight into their behavior. We illustrate by expanding the details for f_1 and f_2. We begin with the case $x = 0.1x_2x_3\cdots$ base 4 so that $x \in [1/4, 1/2]$. According to equation (2.2)

$$f_1(0.1x_2x_3\cdots) = \frac{1}{3} + \frac{1}{3}\left(e^{\frac{i\pi}{3}}\right)f_0(\ell 0.1x_2\cdots).$$

Observe that $f_0(\ell 0.1x_2x_3\cdots) = 0.x_2x_3\cdots$ base 4 so $f_0(\ell x)$ maps $[1/4, 1/2]$ linearly onto the interval $[0, 1]$. Multiplying the interval $[0, 1]$ by $e^{i\pi/3}$ rotates it by an angle of $\pi/3$ radians in the counterclockwise direction producing the complex interval $\left[0, e^{i\pi/3}\right]$ in the plane; multiplying $\left[0, e^{i\pi/3}\right]$ by $1/3$ scales it to the complex interval $\left[0, \frac{1}{3}e^{i\pi/3}\right] = \left[0, \left(\frac{1}{6} + \frac{1}{6}i\sqrt{3}\right)\right]$; and, finally, adding $1/3$ translates this last interval to the complex interval

$$\left[\frac{1}{3}, \frac{1}{3} + \frac{1}{3}e^{i\pi/3}\right] = \left[\left(\frac{1}{3}, 0\right), \left(\frac{1}{2}, \frac{\sqrt{3}}{6}\right)\right].$$

That is, $[1/4, 1/2]$ is mapped linearly by $f_1(x)$ onto an interval of length $1/3$ that starts at the point $1/3 = (1/3, 0)$ and is rotated by an angle of $\pi/3$ radians relative to the x-axis. The endpoint of the interval is $\left(1/2, \sqrt{3}/6\right)$.

Similar analysis reveals that $f_1(.0x_2x_3\cdots)$ maps the interval $[0, 1/4]$ onto $[0, 1/3]$. Notice that the endpoint of $f_1([0, 1/4])$ is the same as the beginning point of the interval $f_1([1/4, 1/2])$. Likewise, $f_1([1/2, 3/4]) = \left[\left(1/2, \sqrt{3}/6\right), (2/3, 0)\right]$ and $f_1([3/4, 1]) = [2/3, 1]$. The final graph of $f_1(I)$, the generator of the Koch curve, is depicted in Figure 2.2.

The second case is $k = 2$ and $x_1 = 1$. According to equation (2.2) and the above analysis

$$f_2(0.1x_2x_3\cdots) = \frac{1}{3} + \frac{1}{3}(e^{\frac{i\pi}{3}})f_1(0.x_2x_3\cdots),$$

so

$$f_2([1/4, 1/2]) = \frac{1}{3} + \frac{1}{3}(e^{\frac{i\pi}{3}})f_1(I).$$

Hence, the interval $[1/4, 1/2]$ is mapped by f_2 onto a rotated (by $e^{i\pi/3}$), scaled (by $1/3$), and translated (by $1/3$) copy of $f_1(I)$ producing the image given in Figure 2.3. The graph of $f_1(I)$ appears as a heavy dashed line. Notice how the graph of $f_2([1/4, 1/2])$ "rests" on one of the four line segments composing the graph of f_1 and is of relative size $1/3$. A complete graph of $f_2(I)$ appears in Figure 2.4. Figure 2.5 isplays the image of f_3.

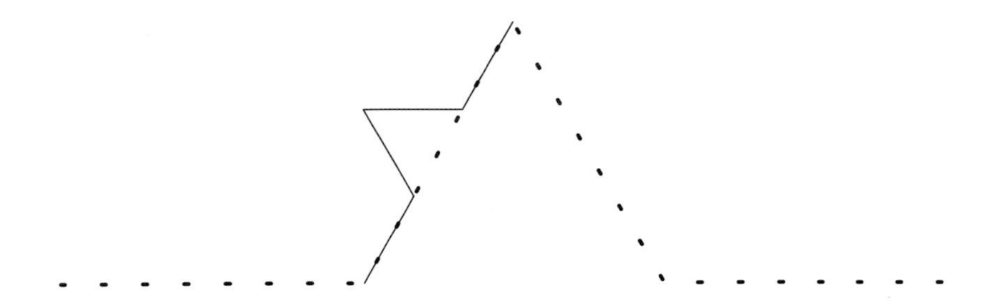

Fig. 2.3: A plot of $f_2([1/4, 1/2])$.

We see that $f_{k+1}(I)$ places a suitably scaled, rotated, and translated copy of the generator onto each line segment forming $f_k(I)$. Consequently, endpoints of any line segment at any stage will always be endpoints of future intervals. Mathematically,

$$f_k\left(\frac{j}{4^k}\right) = f_{k+m}\left(\frac{j}{4^k}\right) \tag{2.3}$$

for any $k, m \in \mathbb{N}$ and $0 \leq j \leq 4^k - 1$.

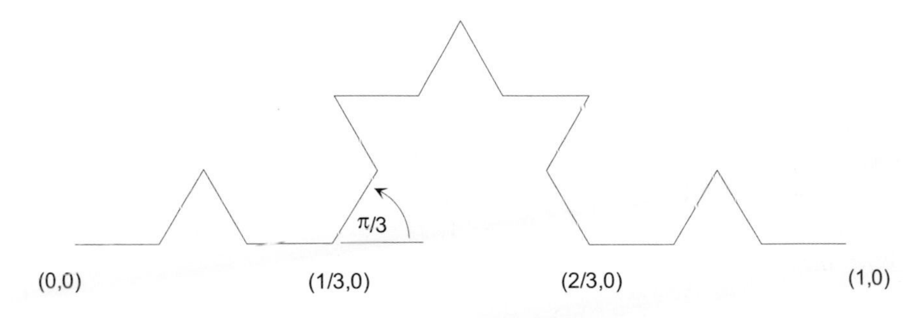

Fig. 2.4: Graph of $f_2(I)$.

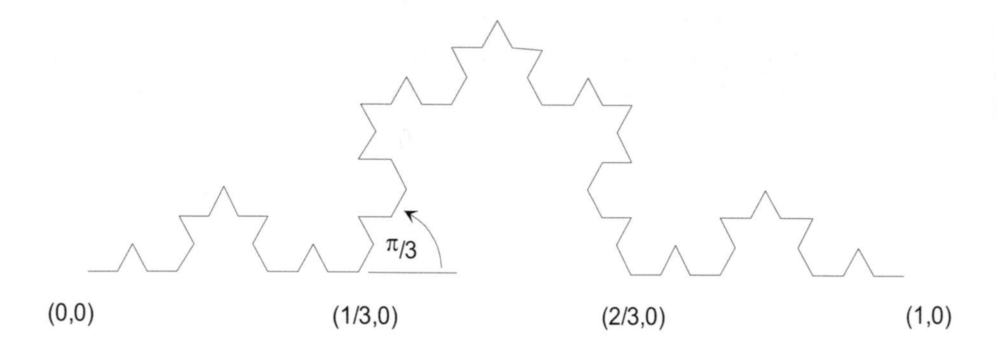

Fig. 2.5: Graph of $f_3(I)$.

2.1.3 *Convergence of the sequence* $\{f_k\}$

We are interested in the convergence of the sequence of functions $\{f_k\}$ from I into $\mathbb{C}$. The distance $\|g - h\|$ between two arbitrary continuous functions, g and h, from I into $\mathbb{C}$ is defined by

$$\|g - h\|_I = \sup\{|g(x) - h(x)| : x \in I\},$$

where $|g(x) - h(x)|$ denotes the modulus of the difference of the complex numbers $g(x)$ and $h(x)$. Recall that for any complex numbers $a, z \in \mathbb{C}$ we have $|az| = |a|\,|z|$, so

$$\|ag\| = |a|\,\|g\|$$

for any function g from I into $\mathbb{C}$.

Since

$$f_{k+1}(x) - f_k(x) = D(x_1)(f_k(\ell x) - f_{k-1}(\ell x)),$$

and $\max\{|D(i)| : i = 0, 1, 2, 3\} = 1/3$ we have the equality

$$\|f_{k+1} - f_k\| = \frac{1}{3}\|f_k - f_{k-1}\|.$$

Hence,

$$\|f_{k+1} - f_k\| = \frac{1}{3}\|f_k - f_{k-1}\| = \cdots = \frac{1}{3^k}\|f_1 - f_0\| < \frac{1}{3^k},$$

since $\|f_1 - f_0\| < 1$.

Applying Theorem C.7 the sequence $\{f_k\}$ is uniformly Cauchy. Hence, by Theorem C.6 and Corollary C.5, $\{f_k\}$ converges uniformly to a continuous function f from I to $\mathbb{C}$. The rate of convergence of $\{f_k\}$ to f is geometric since

$$\begin{aligned}\|f - f_k\| &\leq \sup_{m\geq 1} \|f_{k+m} - f_k\| \qquad (2.4)\\ &\leq \sum_{m=1}^{\infty} \|f_{k+m} - f_{k+m-1}\| \\ &\leq \left(\frac{1}{3}\right)^k \sum_{m=1}^{\infty} \left(\frac{1}{3}\right)^m = \frac{1}{2}\left(\frac{1}{3}\right)^k. \qquad (2.5)\end{aligned}$$

The Koch curve K is the image $f(I)$ of f. Computer generated pictures of K are approximations to K. The inequality (2.5) implies that "If I know the resolution of your image, then I can choose k (the number of iterations) so that you cannot see the difference between $f_k(I)$ and $f(I) = K$". Such images are called *visually acceptable* drawings of the curves.

Observe that by letting $m \to \infty$ in equation (2.3) we have

$$f_k\left(\frac{j}{4^k}\right) = f\left(\frac{j}{4^k}\right)$$

for any $k \in \mathbb{N}$ and $0 \leq j \leq 4^k - 1$.

2.1.4 *An equation for f*

From the recursion equation

$$f_{k+1}(x) = C(x_1) + D(x_1)f_k(\ell x),$$

it follows that

$$f(x) = \lim_{k\to\infty} f_{k+1}(x) = \lim_{k\to\infty} (C(x_1) + D(x_1)f_k(\ell x)) = C(x_1) + D(x_1)f(\ell x).$$

So f satisfies the simple functional equation

$$f(x) = C(x_1) + D(x_1)f(\ell x). \qquad (2.6)$$

According to the preceding formula

$$f(0.1x_2x_3\cdots \text{base}\, 4) = 1/3 + 1/3(e^{\frac{i\pi}{3}})f(0.x_2x_3\cdots \text{base}\, 4).$$

Consequently, the interval $[1/4, 1/2]$ is mapped onto a copy of $f(I)$ of relative size $1/3$. This copy of $f(I)$ starts at the point $1/3 = (1/3, 0)$ and is rotated $\pi/3$ relative to the x-axis. (You should describe the images of $[0, 1/4]$, $[1/2, 3/4]$, and $[3/4, 1]$.) These features of K are seen in Figure 2.1.

2.1.5 *Length of the Koch curve*

We will now demonstrate that the length of the Koch curve is infinite. Let $P = \{t_j\}_{j=0}^{n}$ be a partition of the interval $[a,b]$ and let $h : [a,b] \longrightarrow \mathbb{C}$. Set

$$L(h,P) = \sum_{j=1}^{n} |h(t_j) - h(t_{j-1})|.$$

The *length* of a curve $f : [a,b] \longrightarrow \mathbb{C}$ is defined by

$$\operatorname{len}(f,[a,b]) = \sup\{L(f,P) : P \text{ is a partition of } [a,b]\}.$$

(Section B.4 of Appendix B contains a brief discussion of the length of a curve.) For a fixed positive integer k we know that $f_k(I)$ is composed of 4^k line segments which abut and have length $1/3^k$. This means that $\operatorname{len}(f_k,(I)) = (4/3)^k$, for $k \geq 1$. We note that the curves f and f_k have the same image at each corner point of f_k. If $P = \{t_j\}_{0 \leq j \leq k+1}$ denotes the partition composed of the corner points then

$$\begin{aligned}\operatorname{len}(f,I) &\geq \sum_{j=1}^{k+1} |f(t_j) - f(t_{j-1})| \\ &= \operatorname{len}(f_k, I) \\ &= \left(\frac{4}{3}\right)^k \quad \text{for } k \geq 1.\end{aligned}$$

Since $\lim_{k\to\infty} (4/3)^k = \infty$, the length of the Koch curve is infinite.

If $0 \leq a < b \leq 1$, then $f([a,b])$ contains a reduced size copy of $f(I)$. To see this let $\mu \in \mathbb{N}$ satisfy $\mu > (b-a)^{-1}$ and let ν be the least positive integer such that $a < \frac{\nu}{4^\mu}$. Then $\left[\frac{\nu}{4^\mu}, \frac{\nu+1}{4^\mu}\right] \subset [a,b]$. Also, $f_{\mu+1}\left(\left[\frac{\nu}{4^\mu}, \frac{\nu+1}{4^\mu}\right]\right)$ is a scaled, rotated, and translated copy of $f_\mu(I)$. Hence, $f\left(\left[\frac{\nu}{4^\mu}, \frac{\nu+1}{4^\mu}\right]\right) \subset f([a,b])$ is a scaled, rotated, and translated copy of $f(I)$. Consequently, the length of $f([a,b])$ is also infinite.

2.2 Tangent lines to simple curves in $\mathbb{C}$

In this section we give a definition for a tangent line to a simple curve S at a point p on the curve, and we examine some simple curves, like the Koch curve, that have a tangent line at no point of the curve.

2.2.1 *Definition of a tangent line to a simple curve*

Definition 2.1. The statement that a simple curve S has a **tangent line** at a non-endpoint $p \in S$ means:

(1) S is the image of a one-to-one, continuous map $f : I \to \mathbb{C}$ with $p = f(t)$ and $0 < t < 1$.

(2) Referring to Subsection 1.1.3 and Definitions A.3 and C.9, the limits ($0 < r_u$ if $u \neq t$)

$$\overrightarrow{f_p} = \lim_{u \to t^+} \{\alpha_u : f(u) - f(t) = r_u e^{i\alpha_u}\}$$

and

$$\overleftarrow{f_p} = \lim_{u \to t^-} \{\beta_u : f(u) - f(t) = r_u e^{i\beta_u}\}$$

exist.

(3)

$$\overrightarrow{f_p} = \overleftarrow{f_p} + \pi.$$

Problem 2.1 asks you to show that if g is another such map: $g(I) = f(I)$, with $g(0) = f(0)$, then

$$\overrightarrow{g_p} = \overrightarrow{f_p} \text{ and } \overleftarrow{g_p} = \overleftarrow{f_p}.$$

Definition 2.2. Suppose the simple curve S has a tangent line at the point p and let f be a one-to-one, continuous map from I onto S with $p = f(t), 0 < t < 1$. The tangent line to S at p is the set

$$\{z : z = p + re^{i\overrightarrow{f_p}}, r \in \mathbb{R}\}. \tag{2.7}$$

When S has a tangent line at p, it is said to be **smooth** at p. A curve is a **smooth curve** if it has a tangent line at each point. Problem 2.2 asks you to define smoothness of a simple curve at an endpoint.

Example 2.3. Let $f(t) = e^{it}$, and let $t \in (0, 1)$. We will show that f has a tangent at $p = f(t)$. To this end, suppose $y > 0$ and $t + 2y \in (0, 1)$. Put $x = t + y$; then

$$\begin{aligned} f(x+y) - f(x-y) &= (\cos(x+y), \sin(x+y)) - (\cos(x-y), \sin(x-y)) \\ &= (\cos(x+y) - \cos(x-y), \sin(x+y) - \sin(x-y)) \\ &= (-2\sin y \sin x, 2\sin y \cos x) \\ &= 2\sin y(-\sin x, \cos x) \\ &= 2\sin y(\cos(x+\pi/2), \sin(x+\pi/2)) \\ &= (2\sin y)e^{i(x+\pi/2)} \\ &= (2\sin y)e^{i(t+y+\pi/2)}. \end{aligned}$$

Thus, referring to Subsection 1.1.3 and Definitions A.3 and C.9 again,

$$\begin{aligned}\overrightarrow{f_p} &= \lim_{y \to 0^+} ((t+y) + \pi/2) \\ &= t + \pi/2.\end{aligned}$$

Verify that $\overleftarrow{f_p} = t - \pi/2$. This example continues in Problem 2.3.

2.2.2 *Another construction of the Koch curve*

We describe an alternate construction of the Koch curve that is used to verify that it has no tangent lines. Begin with a triangle E_0 as shown in Figure 2.6. The first stage of the construction removes the equilateral triangle with side length $1/3$ as shown in Figure 2.7. Repeat the construction on the shaded triangles which are similar to the original triangle. The second iteration is shown in Figure 2.8. Note that E_2 is composed of 4 congruent triangles, each with base length $1/3$. The third iteration is shown in Figure 2.9. Each E_k is compact and connected; so $K = \cap_{k \geq 1} E_k$ is compact and connected, and $E_k \to K$ in the Hausdorff distance. (Hausdorff distance is discussed in detail in Chapter 5.) The Koch curve is K. Note that $K \subset E_0$; a copy of K scaled by $1/2$ lies in each of the triangles of E_1; a copy of K scaled by $1/4$ lies in each of the triangles of E_2; etc. This construction of the Koch curve demonstrates visually that the Koch curve is a simple curve.

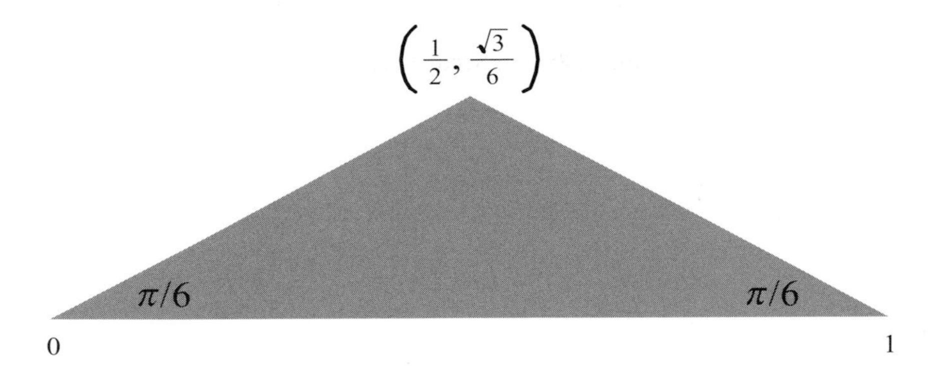

Fig. 2.6: E_0 : Original isosceles triangle.

Because area $E_0 = \sqrt{3}/12$ and area $E_{k+1} = 2/3(\text{area}\, E_k)$, $k = 1, 2, \ldots$ we have area $K = 0$. Also, at each iteration the resulting figure is a collection of suitably scaled, rotated, and translated copies of E_1. We use this construction of

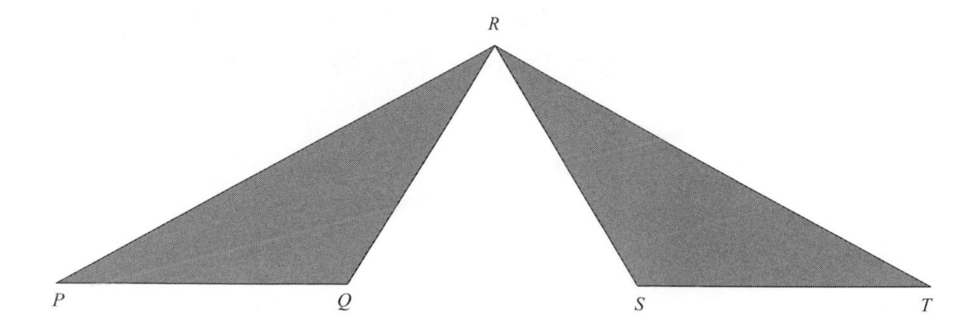

Fig. 2.7: E_1 : First interation in the construction of the Koch curve.

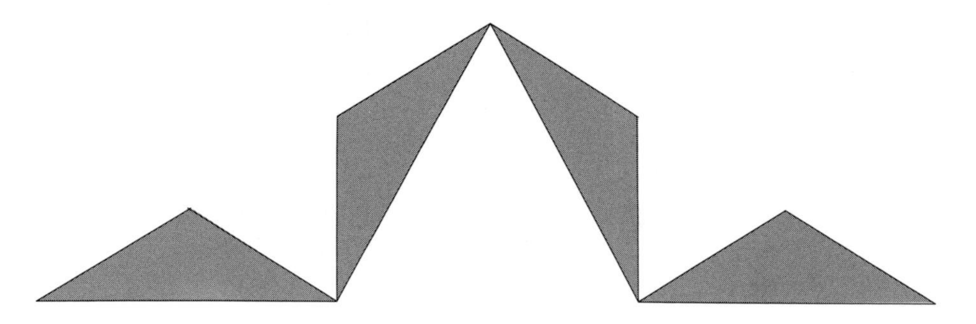

Fig. 2.8: E_2 : Second iteration.

the Koch curve to show that it has a tangent line at no point. Referring to Figure 2.7 and Table 2.1, we have

$$\begin{aligned}
P &= f\,(0.0\,\text{base}\,4) = f\,(0) = C(0), \\
Q &= f\,(0.1\,\text{base}\,4) = f\,(1/4) = C(1), \\
R &= f\,(0.2\,\text{base}\,4) = f\,(1/2) = C(2), \\
S &= f\,(0.3\,\text{base}\,4) = f\,(3/4) = C(3), \text{ and} \\
T &= f\,(1.0\,\text{base}\,4) = f\,(1) = 1.
\end{aligned}$$

The image $f([0,1/2])$ is a subset of triangle PQR, and $f([1/2,1]$ is a subset of triangle RST. For $Z = f(t)$ with $0 \leq t < 1/2$, $Z \neq R$ and $\angle RZS \geq \pi/6$. For $Z = f(t)$ with $1/2 < t \leq 1$, $Z \neq R$ and $\angle RZQ \geq \pi/6$. Finally, for $Z = f(1/2) = R$, $\angle PZQ = \pi/6$.

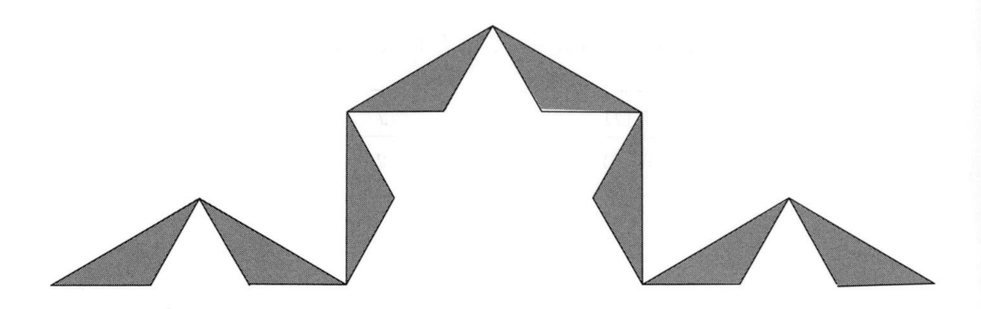

Fig. 2.9: E_3 : Third iteration.

Let $z = f(t) \in K$ and $\epsilon > 0$. The point z is in a set that has the shape of E_1 and base length $< \epsilon$. Thus, there are two points in K on one side of z with distance $< \epsilon$ from z for which the slopes of the cord lines from z to the two points differ by at least $\pi/6$. Consequently, K has no tangent line at z.

2.2.3 *Tangent lines to graphs of continuous maps from I to* $\mathbb{R}$

For a continuous map $g : I \to \mathbb{R}$, the associated graph map $f = f_g$ is defined by $f(t) = (t, g(t))$, $t \in I$. The graph $G = f(I)$ is a simple curve. Moreover, if the curve has a tangent line at a point $p = f(t)$, then $\overrightarrow{f_p} \in [-\pi/2, \pi/2]$.

Example 2.4 (A Graph which has no Tangent Lines). We use base 9 representations of numbers in I for this example: $t = 0.t_1t_2 \cdots t_j \cdots$ base 9, where each t_j is an integer and $0 \le t_j \le 8$ with the usual left shift $\ell t = 0.t_2 \cdots t_j \cdots$ base 9. The initiator f_0 is defined by

$$f_0(t) = \begin{bmatrix} t \\ t \end{bmatrix}.$$

Subsequent functions in the sequence are defined recursively:

$$f_{n+1}(t) = C(t_1) + M(t_1)f_n(\ell t),$$

where the points $C(j)$ given in Table 2.2 and the matrices $M(j)$ below are defined to generate a sequence $\{f_n\}$ of graph maps that converge uniformly to a graph map f:

$$M(j) = \begin{cases} \begin{bmatrix} 1/9 & 0 \\ 0 & 1/3 \end{bmatrix} & \text{if } j = 0, 1, 2, 6, 7, 8 \\ \begin{bmatrix} 1/9 & 0 \\ 0 & -1/3 \end{bmatrix} & \text{if } \quad j = 3, 4, 5 \end{cases}.$$

Table 2.2: Values of $C(j)$ for Example 2.4.

j	$C(j)$	j	$C(j)$	j	$C(j)$
0	$(0,0)$	1	$(1/9,1/3)$	2	$(2/9,2/3)$
3	$(1/3,1)$	4	$(4/9,2/3)$	5	$(5/9,1/3)$
6	$(2/3,0)$	7	$(7/9,1/3)$	8	$(8/9,2/3)$

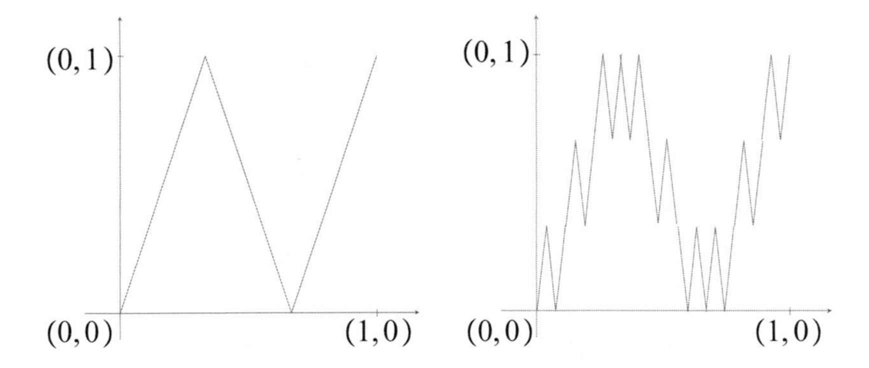

Fig. 2.10: Graphs of f_1 and f_2.

The graphs of $f_1$and f_2 are displayed in Figure 2.10. The points and matrices are organized so that the maps f_n are the graph maps of functions g_n where

$$f_n(t) = (t, g_n(t)).$$

For example,

$$g_1(t) = \begin{cases} 3t, & 0 \leq t \leq 1/3 \\ -3t+2, & 1/3 < t \leq 2/3 \\ 3t-2, & 2/3 < t \leq 1 \end{cases}.$$

Note that

$$|f_{k+1}(t) - f_k(t)| \leq \frac{1}{3} |f_k(\ell t) - f_{k-1}(\ell t)|,$$

$$\|f_{k+1} - f_k\| \leq \frac{1}{3} \|f_k - f_{k-1}\| \leq \cdots \leq \frac{1}{3^k} \|f_1 - f_0\|.$$

Thus the sequence $\{f_k\}$ converges uniformly to f.

Now we need to show that

$$G = \{f(t) = (t, g(t)) : t \in I\}$$

does not have a tangent line at $f(t)$. To this end, fix $t \in I$. The slope function $q = q_t$ defined on $\{u \in I : u \neq t\}$ by

$$q(u) = \frac{g(u) - g(t)}{u - t}$$

is continuous on its domain. Let $n \in \mathbb{N}$. The point $f(t) = (t, g(t))$ is in an n-th level box with lower left corner (a, e) and upper right corner (d, h), where $a = k/9^n \leq t \leq d = (k+1)/9^n$ and $(h - e)/(d - a) = 3^n$. Put $b = k/9^n + 5/9^{n+1}$ and $c = k/9^n + 6/9^{n+1}$. Symmetry permits us to restrict our attention to the case where $a \leq t \leq b$, $g(a) = g(c) = e$ and $g\,(d) = h$. We examine the slope function on $[c, d]$:

$$q(d) = \frac{g(d) - g(t)}{d - t} \geq \frac{g(d) - g(t)}{d - a} \geq 0,$$

$$q(c) = \frac{g(c) - g(t)}{c - t} \leq \frac{g(c) - g(t)}{d - a} \leq 0.$$

Thus,

$$q(d) - q(c) \geq \frac{g(d) - g(t)}{d - a} - \frac{g(c) - g(t)}{d - a} = \frac{g(d) - g(c)}{d - a} = \frac{h - e}{d - a} = 3^n.$$

Consequently, $|q(u)|$ takes every value between 0 and $3^n/2$ on the interval $[c, d] \subset [t - 1/9^n, t + 1/9^n] \cap I$. We are done.

Example 2.5. The Cantor function ϕ is a continuous, nondecreasing map of I onto I that maps the standard Cantor set C onto I. Focusing on Figure 1.14, we give two descriptions of ϕ. For our first description of ϕ notice that C corresponds to the base of Figure 1.14 and I corresponds to the top. Points $p \in C$ have a base 3 expansion composed exclusively of 0's and 2's: $p = 0.(2t_1)(2t_2)\cdots(2t_n)\cdots$ base 3, where each t_n is either 0 or 1. We first define ϕ on C to be the map that takes such a p to

$$q = .(t_1)(t_2)\cdots(t_n)\ \text{base}\ 2.$$

This map is a continuous, nondecreasing map of C onto I that takes the two endpoints of a removed segment onto the same point in I; for example, both points $1/3 = .02222\cdots$ base 3 and $2/3 = .20000\cdots$ base 3 map onto the point $1/2 = .01111\cdots$ base 2 $= .10000\cdots$ base 2. The Cantor function ϕ is the nondecreasing extension of this map to I, so ϕ maps the interval $[1/3, 2/3]$ to the point $1/2$.

We will need the concept of density in the second description of the Cantor function.

Definition 2.3. A subset E of S is **dense** in S if any point of S is a limit point (See Definition A.3) of points in E.

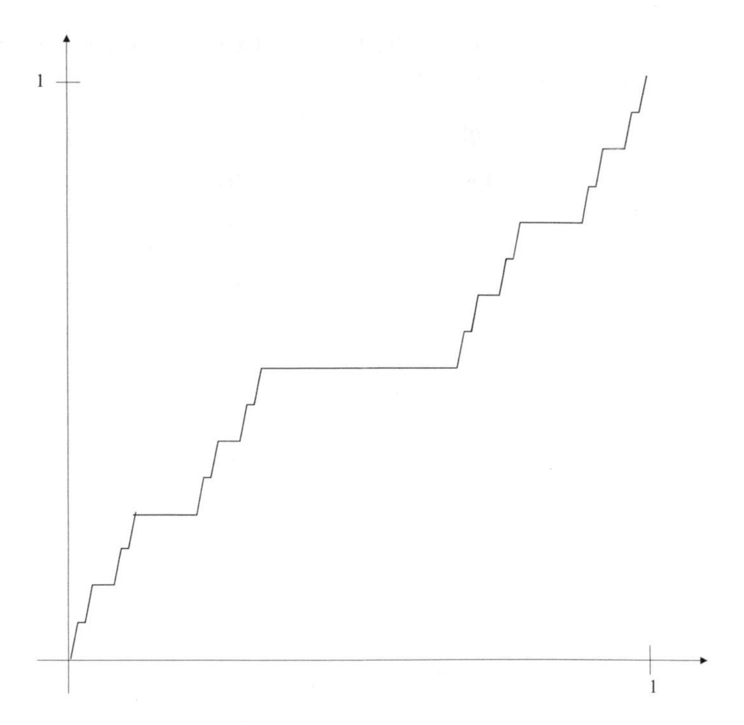

Fig. 2.11: An approximation to the graph of the Cantor function.

The second description of ϕ proceeds in the following way. The Cantor set is what is not removed from the base edge I of the unit square in Example 1.5, and the top edge of the unit square in the figure is a copy of I. The triangle removed at Step 1 identifies the interval $[1/3, 2/3]$ with the point $1/2$. The 2 triangles removed at Step 2 identify the interval $[1/9, 2/9]$ with the point $1/4$ and the interval $[7/9, 8/9]$ with the point $3/4$. The 2^2 triangles removed at Step 3 identify the interval $[1/27, 2/27]$ with $1/8$, the interval $[7/27, 8/27]$ with $3/8$, the interval $[19/27, 20/27]$ with $5/8$ and the interval $[25/27, 26/27]$ with $7/8$. Step n removes 2^{n-1} triangles and identifies 2^{n-1} intervals on the base with 2^{n-1} points $(1/2^n, 3/2^n, \dots, (2^n - 1)/2^n)$ on the top. The Cantor function ϕ maps 0 to 0, 1 to 1, and each of the intervals to its associated point:

$$\begin{aligned} \phi(0) = 0, \phi(1) = 1,& \\ \phi([1/3, 2/3]) = 1/2,& \\ \phi([1/9, 2/9]) = 1/4,\ & \phi([7/9, 8/9]) = 3/4, \\ \phi([1/27, 2/27]) = 1/8,\ & \text{etc.} \end{aligned}$$

It remains to define ϕ at points in the Cantor set that are not 0, 1, or endpoints of removed segments. We can easily do this since the current domain of the Cantor function is dense in I and the current range is the set

$$\{j/2^n : 0 \leq j \leq 2^n, n \in \mathbb{N}\},$$

which is dense in I. For $x \in I$ with $\phi(x)$ not defined above, put

$$\begin{aligned}\phi(x) &= \sup\{\phi(y) : y < x \text{ and } \phi(y) \text{ is defined}\} \\ &= \inf\{\phi(y) : x < y \text{ and } \phi(y) \text{ is defined}\}.\end{aligned}$$

2.2.3.1 *Modified Cantor functions*

We can modify the Cantor function to a map from I to the plane by stretching each base interval around the corresponding triangle instead of mapping it to the apex of the triangle. You should show that this modification is discontinuous at all points in the Cantor set and continuous at all other points in I.

Another modification is constructed by relaxing the stretching as follows. The endpoints of an interval don't change, but an interval of length 3^{-n} is mapped onto a triangle of height $2^{-(n-1)}$ as shown in Figure 3.8. You should verify that this modification of ϕ, which we name Φ, is a continuous map from I to the plane.

2.2.3.2 *The graph G_ϕ of the Cantor function*

An approximation to the graph G_ϕ of ϕ appears in Figure 2.11. The following project asks you to verify some properties of this graph.

Project.

(a) Show that the length of G_ϕ is equal to 2.
(b) Show that G_ϕ has a horizontal tangent line at each point $(x, \phi(x))$, where $x \notin C$.
(c) Show that if G_ϕ has a tangent line at a point $(x, \phi(x))$, where $x \in C$, then the tangent line is a vertical line.
(d) Show that G_ϕ cannot have a tangent line at every point of I.

In Chapter 7 we will explain why G_ϕ has a vertical tangent line at most of the points in C, where "most" has a precise meaning. The graph of the Cantor function is sometimes called "The Devil's Staircase". This vilification contrasts vividly with our name "The Magic Wand" for the image $\phi(I)$ of the Cantor function. At this point $\phi(I)$ is merely the interval I, but it is ready to be transformed in Chapter 3 into curves like the image of Φ.

2.3 Problems

PROBLEM 2.1. Referring to Definition 2.1, show that if g is another such map with $g(0) = f(0)$, $g(s) = f(t) = p$, and $g(1) = f(1)$, then

$$\overrightarrow{g}_p = \overrightarrow{f}_p \text{ and } \overleftarrow{g}_p = \overleftarrow{f}_p.$$

PROBLEM 2.2. Define smoothness of a simple curve at an endpoint of the curve.

PROBLEM 2.3. Given a function $g : [a, b] \to \mathbb{R}$ and a point $t \in (a, b)$; the **derivative** g of at t is defined

$$g'(t) = \frac{dg}{dt}(t) = \lim_{y \to 0} \frac{g(t + y) - g(t)}{y}$$

when this limit exists as a number in $\mathbb{R}$.

(1) Draw the first quadrant arc of the unit circle from $(1, 0)$ to $(0, 1)$. Let $t \in (0, \pi/2)$. Draw the ray from $(0, 0)$ through $(\cos t, \sin t)$ and $(1, \tan t)$. Explain why this ray is smooth, has angle t at each point, and is perpendicular to the interval $[(\cos(t-u), \sin(t-u)), (\cos(t+u), \sin(t+u)]$ if $0 < t - u < t + u < 1$. Draw the vertical intervals $[(\cos t, 0), (\cos t, \sin t)]$ and $[(1, 0), (1, \tan t)]$. Notice that t is equal to the length of the circular arc from $(1, 0)$ to $(\cos t, \sin t)$. Show (Section B.4 and elementary geometry on your drawing) that $0 < \sin t < t < \tan t$. Explain why $\lim_{t \to 0^+} \sin t = 0$, $\lim_{t \to 0^+} \cos t = 1$,and $\lim_{t \to 0^+} \frac{\sin t}{t} = 1$.

(2) Explain why Example 2.3 implies that, for $t \in (0, 1)$,

$$\frac{d \cos}{dt}(t) = -\sin t$$

and

$$\frac{d \sin}{dt}(t) = \cos t.$$

(3) Put $C(t) = (t, \cos t)$ and $S(t) = (t, \sin t)$. Show that these graphs have tangent lines at the images of points $t \in (0, 1)$. Describe the tangent lines when $t = \pi/4$.

PROBLEM 2.4. Let S be the graph of a continuous map g from I to $\mathbb{R}$. Show that S has a non-vertical tangent line at a point $(t, g(t))$ with $0 < t < 1$ if, and only if,

g is differentiable at the point $t \in (0,1)$. According to this exercise, the function defined in Example 2.4 is nowhere differentiable.

PROBLEM 2.5. Define $f : I \to \mathbb{R}$ by $g(0) = 0$ and $g(t) = t^2 \sin(1/t^2)$ if $t > 0$ and let $\overrightarrow{f}(t) = (t, g(t))$ for $t \in I$.

(a) Show that $\overrightarrow{f}_0 = 0$.

(b) Show that if $t > 0$ and $p_t = (t, f(t))$, then $\overrightarrow{f}_{p_t} = \arctan(2t\sin(1/t^2) - (2/t)\cos(1/t^2))$. Thus, the graph of f is a smooth curve.

(c) Show that if $0 < w < 1$, then $\{\tan \overrightarrow{f}_{p_t} : t \in [0, w]\} = \mathbb{R}$.

PROBLEM 2.6. Sketch the simple generalized curves defined for $t \geq 0$ below, and discuss their smoothness properties.

(a) $a(t) = te^{it}$.

(b) $b(0) = 0$ and $b(t) = te^{i/t}$ otherwise.

(c) $c(0) = 0$ and $c(t) = t^2 e^{i/t}$ otherwise.

(d) A circular wheel of radius 1 starts with its center at $(0,0)$ and rolls along the line $y = -1$ in $\mathbb{R}^2$. The position of the center of the wheel at time t is $(t, 0)$. The generalized curve $d(t)$ represents the position of the point on the wheel that starts at $(1,0)$.

(e) The generalized curve $g(t)$ represents the position of the point on the wheel described in the previous part that starts at $(1/2, 0)$.

PROBLEM 2.7. The statement that a point p on a simple curve K is a **spiral point** of K means that every half line in $\mathbb{R}^2$ emanating from p intersects K in an infinite set. For example, $0 = (0,0)$ is a spiral point of the simple curve b in the preceding problem. Denote the set of spiral points of a simple curve K by $SP(K)$.

(a) Construct a simple curve $K = m(I)$ with $m(0) = 0$, $m(1) = 1 = (1,0)$, and $SP(K) = \{1/2\} = \{(1/2, 0)\}$.

(b) Construct a simple curve $K = m(I)$ with $m(0) = 0$, $m(1/n) = 1/n, n \in \mathbb{N}$, and $SP(K) = \{0\} \cup \{1/n : n \in \mathbb{N}\}$.

(c) Determine whether or not you can construct a simple curve $K = m(I)$ with $m(0) = 0$, $m(1) = 1$, and $SP(K)$ equal to the Cantor set. If you can, describe the map m heuristically (without equations) - if at first you do not succeed, try again.

Chapter 3

Curves and Cantor sets

Cantor sets and curves are very different compact sets. A Cantor set is closed, every point of a Cantor set is a limit point of it, and every neighborhood of a point contains a subset that is a Cantor set. Cantor sets are prevalent in I and in the unit square U. Curves are connected, and Cantor sets are totally disconnected. So, how do Cantor sets and curves relate? A continuous map defined on a closed subset of I can be extended to a continuous map on I. Thus, we can construct curves with desired properties by constructing continuous maps defined on Cantor sets in I and then extending the maps to I. This technique is especially useful for constructing simple curves with interesting properties.

Several examples of unusual curves are presented in this chapter, among them is a construction using Cantor sets which produces a simple curve that has positive area at each point! (See Subsection 3.2.2.) We show that any two points in a curve M are the endpoints of a simple curve that is contained in M. We also consider examples of generalized curves, and we show that every compact set is a continuous image of a Cantor set. Finally, we introduce local connectedness and conclude by presenting the Hahn-Mazurkiewicz Theorem, which we will apply in a subsequent chapter to show that a connected fixed set of an Iterated Function System is a curve.

3.1 A square is a curve!

Set $U = [0,1] \times [0,1]$. We use base 9 for this example: $0 \leq t_j \leq 8$, with

$$t = 0.t_1 t_2 t_3 \cdots = \sum_{j \geq 1} \frac{t_j}{9^j} \in I$$

and $lt = 0.t_2 t_3 \cdots$.

We define an initiator $f_0(t) = (t, t)$, the line segment in U connecting the

Table 3.1: Rotation and translation factors for mapping I onto U.

j	0	1	2	3	4	5	6	7	8
3C(j)	$(0,0)$	$(1,1)$	$(0,2)$	$(1,3)$	$(2,2)$	$(1,1)$	$(2,0)$	$(3,1)$	$(2,2)$
3V(j)	1	i	1	$-i$	-1	$-i$	1	i	1

points $(0,0)$ and $(1,1)$. For $n \geq 1$, the functions f_n are defined recursively by the formula

$$f_n(t) = C(t_1) + V(t_1) f_{n-1}(lt)$$

where the complex numbers $C(j)$ and $V(j)$ are defined in Table 3.1 where we list the values of $3C(j)$ and $3D(j)$. For example, multiplication by $V(1) = i/3$ scales a set by $1/3$ and then rotates it by an angle of $\pi/2$ about $(0,0)$ in a counterclockwise direction.

In particular,

$$f_1([1/9, 2/9]) = [(1/3, 1/3), (0, 2/3)].$$

The image $f_1(I)$ consists of similar rotations, scalings, and translations of $f_0(I) = [(0,0), (1,1)]$. Figure 3.1 displays $f_1(I)$ as well as three other rotations of $f_1(I)$ we will need.

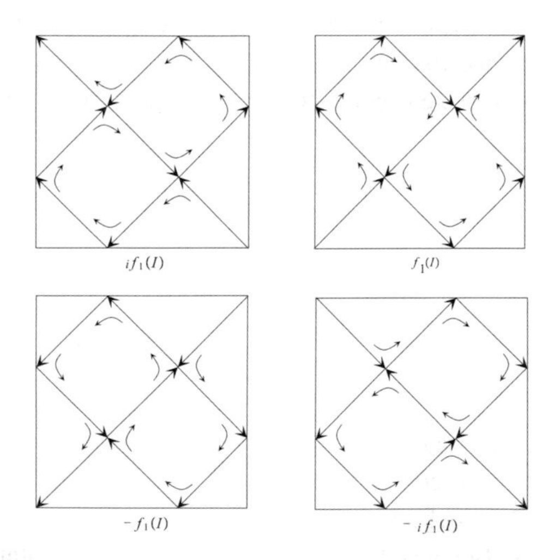

Fig. 3.1: Graphs of $f_1(I)$ and the rotations $if_1(I)$, $-f_1(I)$, and $-if_1(I)$.

Table 3.2: Rotations and translations to produce $f_1(I)$.

$\frac{1}{3}((0,2)+f_0(I))$	$\frac{1}{3}((1,3)-if_0(I))$	$\frac{1}{3}((2,2)+f_0(I))$
$\frac{1}{3}((1,1)+if_0(I))$	$\frac{1}{3}((2,2)-f_0(I))$	$\frac{1}{3}((3,1)+if_0(I))$
$\frac{1}{3}((0,0)+f_0(I))$	$\frac{1}{3}((1,1)-if_0(I))$	$\frac{1}{3}((2,0)+f_0(I))$

According to Table 3.2 there are three rotations and nine translations, including the $(0,0) \longrightarrow (0,0)$ translation, needed to produce Figure 3.1. Now applying these same rotations and translations with a scaling of $1/3$ to $f_1(I)$ produces $f_2(I)$ which is depicted in Figure 3.2.

The convergence argument that we used for the Koch curve in Chapter 2 (See subsection 2.1.3.) applies here. The equality

$$\|f_{n+1} - f_n\| = (1/3)\|f_n - f_{n-1}\|$$

permits us to conclude that the sequence $\{f_n\}$ converges uniformly to a continuous map f of I into $\mathbb{C}$. Notice that the map f satisfies the functional equation

$$f(t) = C(t_1) + V(t_1)f(lt).$$

It remains to show that $f(I) = U$. Clearly, $f(I) \subset U$. To show that $U \subset f(I)$, we let $z \in U$ and find $t \in I$ for which $f(t) = z$. A point $z \in U$ is in at least one of the nine closed sub-squares U_{t_j}, $0 \leq t_j \leq 8$, displayed in Figure 3.3 which comprise U. Each of these nine sub-squares is composed of nine sub-squares $U_{t_1 t_2}$ which are oriented to conform to f_2. (Again, see Figure 3.3.) Continuing this process, there is a nested sequence $\{U_{t_1 t_2 \cdots t_n}\}_{n \geq 1}$ of closed sub-squares for which

$$z = U_{t_1 t_2} \cdots = \bigcap_{n \geq 1} U_{t_1 t_2 \cdots t_n}$$

and

$$f(0.t_1 t_2 \cdots t_n) = f_n(0.t_1 t_2 \cdots t_n) \in U_{t_1 t_2 \cdots t_n}.$$

Thus,

$$f(0.t_1 t_2 \cdots) = z.$$

We illustrate this strategy by finding the $t \in I$ which maps onto $z = (1/4, 1/2)$. The x and y coordinates of z are easily written in base 3 as $1/4 = 0.\underline{02}$ base 3 and $1/2 = 0.\underline{1}$ base 3 $= 0.\underline{11}$ base 3. Thus, referring to Figure 3.3, $z \in U_1$ and $z \in U_{15}$. Because U and U_{15} have the same orientation and both $1/4$ and $1/2$ have period two base 3 representations, we can conclude that

$$f(0.\underline{15}) = f(7/40) = (1/4, 1/2).$$

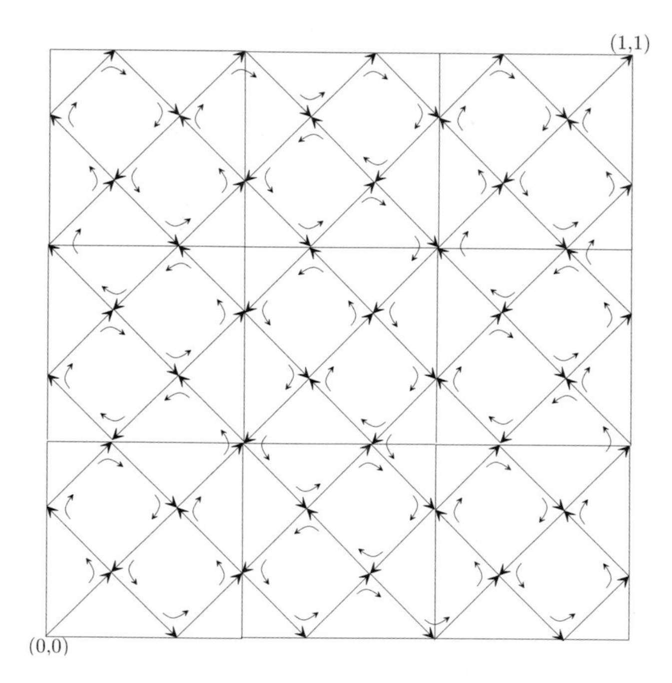

Fig. 3.2: The graph of $f_2(I)$ traces the straight arrows beginning at $(0,0)$ and terminating at $(1,1)$ following the route determined by the curved arrows.

3.2 Simple curves

We have defined a simple curve to be a one-to-one, continuous image of I. We have seen in Subsection 2.1.4 that the Koch curve is a simple curve. According to Section C.7 in Appendix C, a simple curve is homeomorphic to I.

The following project encourages you to entertain some geometric properties of curves.

Project. The continuous image of a compact, connected set in the plane is a compact, connected set in the plane (Proposition C.13). A compact, connected set on a line is an interval (Proposition C.10). If f is a one-to-one, continuous map from a compact set $S \subset \mathbb{R}^2$ onto $T \subset \mathbb{R}^2$ then the inverse map f^{-1} is a one-to-one, continuous map from T onto S (Proposition C.16).

(a) Show that the unit circle $\{z : |z| = 1\}$ is not homeomorphic to I.
(b) Show that the Koch snowflake is homeomorphic to the unit circle.
(c) Show that the Koch snowflake is not homeomorphic to I.
(d) Show that the Koch snowflake is not a self-similar curve.

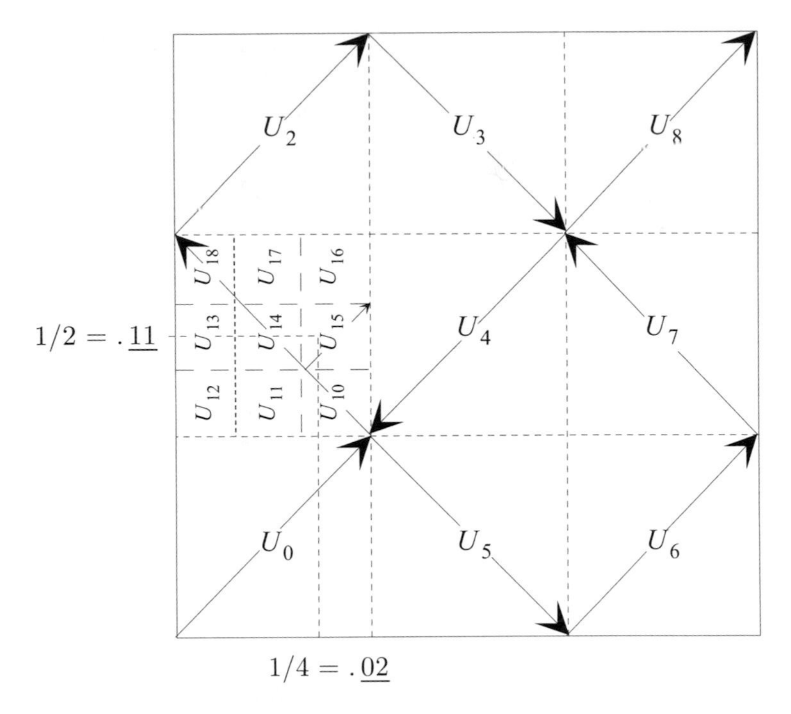

Fig. 3.3: Orientation.

(e) Show that the unit square $U = \{z = (x, y) : 0 \leq x \leq 1, 0 \leq y \leq 1\}$ is not homeomorphic to I.
(f) Show that if a curve contains a homeomorphic image of the unit circle, then it is not a simple curve.

Propositions 3.1 and 3.2 below show that if p and q are two points in a curve M, then M contains a simple curve K from p to q. This result applies to some surprisingly complicated curves. For example, visually find a simple curve from $p = e^{i\sqrt{2}}$ to $q = 2e^{i\sqrt{3}}$ in the curve depicted in Figure 3.12. We will prove the following two propositions after presenting some *not so simple* simple curves.

Proposition 3.1. *Let p and q be two points in a curve $M = f(I)$. Then M contains a curve $g(I)$, where the continuous map g satisfies three conditions: (1) $g(0) = p$, (2) $g(1) = q$, and (3) for each $t \in g(I)$, $g^{-1}(t)$ is an interval in I.*

Proposition 3.2. *Suppose f is a continuous map that satisfies the three conditions described in Proposition 3.1 above. Then there is a one-to-one continuous map h with $h(0) = p$, $h(1) = q$ and $h(I) = f(I)$.*

3.2.1 *A homeomorphism g with* $C \times C \subset g(I)$

Denote the standard, middle-thirds-removed Cantor set by C. We develop an explicit definition of a function g to construct a simple curve $g(I)$ in U with the property that $C \times C \subset g(I)$ and the area of $g(I)$ is equal to zero.

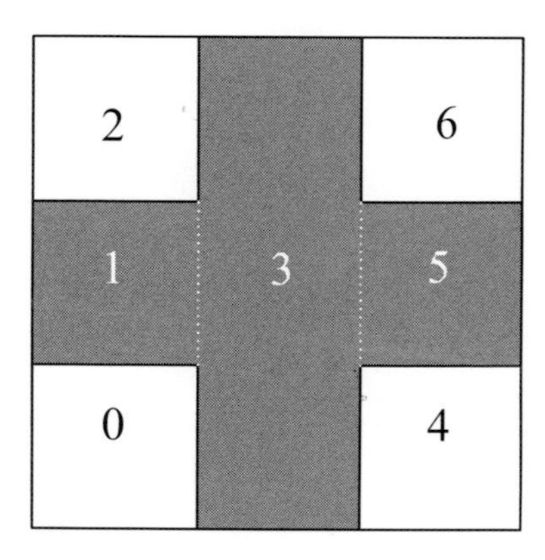

Fig. 3.4: The sets U_k.

Our map g will weave through the seven rectangles displayed in Figure 3.4, passing diagonally across the odd numbered rectangles and generating reduced copies of the full map on each of even numbered squares; g will map the interval $[j/7, (j+1)/7]$ into rectangle j, where $j = 0, \ldots, 6$.

For this construction we use base 7 : for $0 \leq t_j \leq 6$,

$$t = 0.t_1t_2t_3 \cdots \text{base } 7 = \sum_{j \geq 1} \frac{t_j}{7^j} \in I$$

so I is partitioned into 7 similar sub-intervals

$$I_k = [0.k \text{ base } 7, 0.(k+1) \text{ base } 7] = [k/7, (k+1)/7], 0 \leq k \leq 6.$$

Referring back to Figure 1.8, we combine squares 3, 4 and 5 into one rectangle and label the resulting 7 rectangles displayed in Figure 3.4 as $U_k, 0 \leq k \leq 6$. Each image $g(I_k)$ will be a subset of the corresponding rectangle U_k. For k odd, g is defined linearly on I_k by the solid arrow that appears in U_k in Figure 3.5a.

For $n = 1$, the 4 rectangles U_{k_1}, with k_1 even, are squares whose union contains $C \times C$. Each of these even numbered squares is the union of 7 rectangles $U_{k_1k_2}$ as displayed in Figure 3.4. For k_2 odd, g is defined linearly on $I_{k_1k_2}$ and its image is displayed by the solid arrow that appears in $U_{k_1k_2}$ in Figure 3.5b.

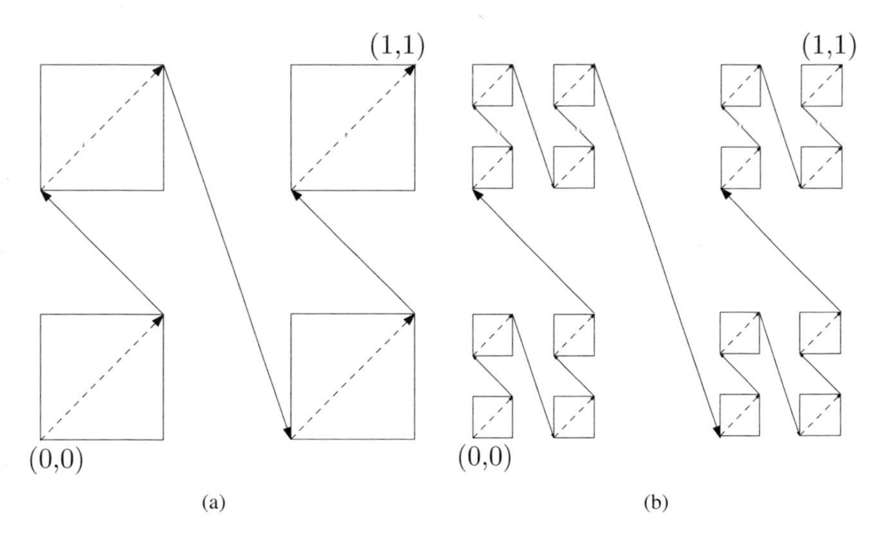

Fig. 3.5: Portions of the graph of g. The figure on the left represents the image of g on U_1, U_3, and U_5. The figure on the right is the image of g after the second partitioning of I.

For $n = 2$, the 4^2 rectangles $U_{k_1 k_2}$, with both k_1 and k_2 even, are squares whose union contains $C \times C$. We continue this process on each $U_{k_1 k_2}$.

After continuing for $n = 3, 4, \ldots$, the function g is defined for all points $t = 0.t_1 t_2 \cdots$ for which at least one t_j is odd. The points with all t_j even map onto $C \times C$ as follows

$$g(0.t_1 t_2 t_3 \cdots) = \bigcap_{n \geq 1} U_{t_1 t_2 t_3 \cdots t_n}.$$

Thus, g is defined on I and $g(I) \supset C \times C$. To show that g is a one-to-one map, refer to Figure 3.5 and observe that if $0 \leq t < u \leq 1$, then there is an interval of the form $[j/7^n, (j+1)/7^n]$ in the segment (t, u).

Remark 3.1. Using Table 3.3 we have explicit formulas for $g(t)$. If t has all even entries, then

$$g\left(0.t_1 t_2 t_3 \cdots t_n\right) = \sum_{j=1}^{n} \frac{H(t_j)}{3^{j-1}},$$

and

$$g(t) = \lim_n g\left(0.t_1 t_2 t_3 \cdots t_n\right) = \sum_{j=1}^{\infty} \frac{H(t_j)}{3^{j-1}}.$$

Table 3.3: Values in construction of g.

j	$H(j)$	$V(j)$
0	$(0,0)$	–
1	$(1/3,1/3)$	$(-\frac{1}{3},\frac{1}{3})$
2	$(0,2/3)$	–
3	$(1/3,1)$	$(\frac{1}{3},-1)$
4	$(2/3,0)$	–
5	$(1,1/3)$	$(-\frac{1}{3},\frac{1}{3})$
6	$(2/3,2/3)$	–

If $j > 1$ and t_j is the first odd entry in t, then

$$g(t) = H(t_1) + \frac{1}{3}H(t_2) + \frac{1}{3^2}H(t_3) + \cdots + \frac{1}{3^{j-1}}H(t_j) + \frac{1}{3^{j-1}}V(t_j)(l^{(j)}t).$$

Refer to Chapter 1 (Subsection 1.3.2) and note that for $0 \leq h < 1$, we can change the lengths of the sides appropriately in the construction above and obtain a corresponding continuous map g_h with $g_h(I) \supset C_h \times C_h$.

3.2.2 *Simple curves with positive area*

The concept of areas of subsets of $\mathbb{C}$ is discussed in Appendix B. You may need to consult some details there to verify the statements in the following.

Let $U = [0,1] \times [0,1]$. We will verify that $\text{area}(C_h \times C_h) = h^2$ by calculating $\text{area}\,(U - C_h \times C_h)$ and using the formula

$$\text{area}(C_h \times C_h) = 1 - \text{area}\,(U - C_h \times C_h)\,.$$

Fix h and set $B_h = [(0,h)\,,(1,h)]$. Recall that $t = 1 - h$ and that S_n denotes the subset of U that remains after *Step* n in Example 1.5. Let T_n denote $S_n \cap B_h$. Then

$$T_1 \supset T_2 \supset \cdots$$

and

$$C_h = \bigcap_{n \geq 1} T_n.$$

Put $U_n = B_h - T_n$ and for convenience set $r_n = \text{len}(U_n)$. Then

$$r_n = t\left(\frac{1}{3} + \frac{2}{3^2} + \cdots + \frac{2^{n-1}}{3^n}\right) = t\left(1 - \left(\frac{2}{3}\right)^n\right)$$

is the total length of the segments removed from B_h in the first n steps. Since T_n is composed of 2^n closed intervals, $T_n \times T_n$ is composed of 4^n pairwise disjoint closed squares. Also,

$$\text{area}(T_n \times T_n) = (1 - r_n)^2$$

and

$$C_h \times C_h = \bigcap_{n \geq 1} (T_n \times T_n).$$

For each positive integer n set $V_n = U - (T_n \times T_n)$, so that the area of V_n is $1 - (1 - r_n)^2$. Set $V = \cup_{n \geq 1} V_n$. The set V_1 (Figure 3.4) is composed of 3 nonoverlapping rectangles; V_2 is composed of the rectangles in V_1 plus $3 \cdot 4$ new rectangles, 3 in each of the 4 squares in $T_1 \times T_1$. Continuing, V_{n+1} is composed of the rectangles in V_n plus $3 \cdot 4^n$ new rectangles, 3 in each of the 4^n squares in $T_n \times T_n$.

Consequently,

$$\begin{aligned} \text{area}\,(C_h \times C_h) &= 1 - \text{area}\,(V) \\ &= 1 - \lim_n \text{area}\,(V_n) \\ &= \lim_n \,(1 - r_n)^2 = h^2. \end{aligned}$$

The set $C_h \times C_h$ has area equal to h^2 and the area of an interval in $\mathbb{R}^2$ is equal to 0. Thus, you can show (Problem 3.12) that the area of $g_h(I)$ is equal to h^2. Problem 3.16 asks you to show that the curve $g_h(I)$ has a tangent line at a point $g_h(p)$ if and only if $p \notin C_h$. According to this problem for a given $s \in [0, 1)$ we have a simple curve S in the unit square for which the area of the set of points at which S has no tangent lines is equal to s.

Later we will show that any compact set K in $\mathbb{R}^2$ is a continuous image of C. Thus, every compact set K in $\mathbb{R}^2$ can be embedded in a curve with the same area.

Definition 3.1. The statement that a curve K has **positive area at a point** $p \in K$ means that if $r > 0$, then

$$\text{area}\,(K \cap \{y \in \mathbb{C} : |y - p| \leq r\}) > 0.$$

Note that the curve $g_h(I)$ has positive area at each point of $C_h \times C_h$.

Example 3.1. In this example we construct a simple curve $K = f(I)$ that has positive area at each point on the curve. The construction of f is a modification of that for g_h. We need to ensure that there is positive area at the points in the odd rectangles. Although the details are a bit complex, the level by level transitions are clear. We will use two procedures to choose 7 sub-rectangles of a rectangle. The first (*even parity*) decomposition is the one we previously used to partition a rectangle into 7 sub-rectangles. (See Figure 3.4.) The second (*odd parity*) decomposition of a rectangle uses only 7 rectangles in a partition of the rectangle into 49 similar sub-rectangles. Let rectangle $U_{k_1 k_2 \cdots k_n} = [s,t] \times [u,v]$. For an odd parity decomposition, we put, for $j = 0, \ldots, 6$,

$$U_{k_1 k_2 \cdots k_n j} = \left\{ (x,y) : \left\{ \begin{array}{c} \frac{7-j}{7}s + \frac{j}{7}t \le x \le \frac{6-j}{7}s + \frac{j+1}{7}t \\ \frac{7-j}{7}u + \frac{j}{7}v \le y \le \frac{6-j}{7}u + \frac{j+1}{7}v \end{array} \right\} \right\}.$$

These are the seven similar sub-rectangles that contain the diagonal of $[s,t] \times [u,v]$. An even parity decomposition keeps even parity on sub-rectangles 0, 2, 4 and 6; it assigns odd parity to 1, 3 and 5. An odd parity decomposition keeps odd parity on sub-rectangles 0, 2, 4 and 6; it assigns even parity to 1, 3 and 5. Note that the parity changes at level n when

$$P_n(t) = \sum_{j \le n} t_j \pmod 2$$

changes value: $P_n(t) = 0$ corresponds to even parity and $P_n(t) = 1$ corresponds to odd parity. If we map U onto $[0,a] \times [0,b]$ via the map

$$(x,y) \to (ax, by),$$

the area h^2 of $C_h \times C_h$ maps to a set with area abh^2.

We begin with an initial partition of U into 7 sub-rectangles $U_0, \ldots, U_6$ as in the construction for $C \times C$, except with width h. At level 1, we begin the construction of a $1/3$-sized copy of $C_h \times C_h$ in each of the even parity rectangles U_0, U_2, U_4 and U_6 by partitioning each of them into 7 sub-rectangles of the appropriate size. In each of the odd parity rectangles U_1, U_3, U_5, we label the 7 diagonal rectangles as shown in Figure 3.6.

We go from an odd parity rectangle to an even parity rectangle $U_{k_1 \cdots k_n}$ when k_n is odd: $k_n = 1, 3$ or 5. When this occurs and $U_{k_1 \cdots k_n}$ has width a and height b, an $a \times b$-sized copy of $C_h \times C_h$ is constructed in $U_{k_1 \cdots k_n}$. This copy of $C_h \times C_h$ is generated by sequences corresponding to

$$t = 0.k_1 \cdots k_n \cdots k_m \cdots,$$

where k_m is even for all $m > n$. Accordingly, $f(t)$ is a assigned a value associated with the value of $g_h(t)$. When k_n is even, parity changes occur for some sub-rectangles $U_{k_1 \cdots k_n \cdots k_m}$.

At this point, $f(t)$ is defined at all points $t \in I$ whose base 7 entries t_m are all even after a finite number of terms. These points are dense in I and f

is uniformly continuous on its current domain. We can extend f to I with the requisite properties by putting

$$f(t) = f(0.k_1k_2\cdots) = \cap_{n=1}^{\infty} U_{k_1\cdots k_n}.$$

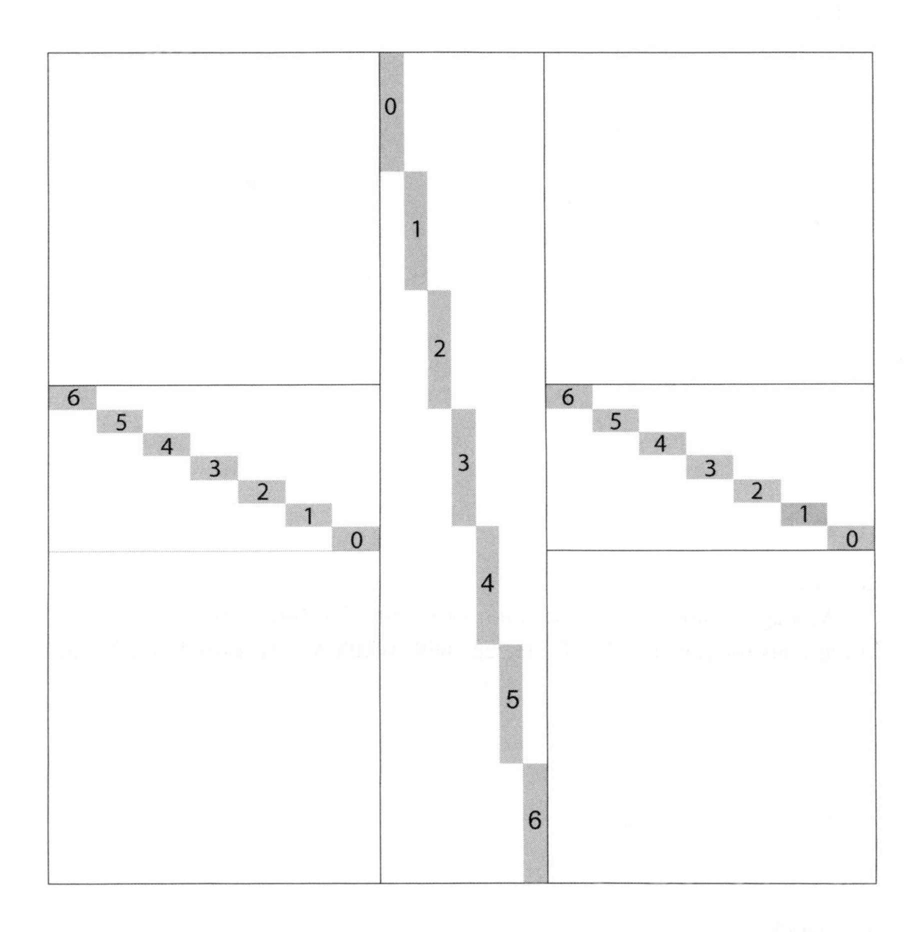

Fig. 3.6: Odd parity.

Definition 3.2. A point p of a curve K is a **point of density 1** if

$$\lim_{r\to 0^+} \frac{\text{area}(K \cap \{y \in \mathbb{C} : d(p,y) \leq r\})}{\pi r^2} = 1.$$

In Problem 3.2 you will be asked to construct a simple curve with a point of density 1.

3.2.3 *Proofs of Propositions 3.1 and 3.2*

We now present the proofs of the propositions given earlier.

3.2.3.1 *Proof of Proposition 3.1*

Without loss of generality, let f be a continuous map from I to $\mathbb{C}$ with $p = f(0) \neq f(1) = q$ and set $K = f(I) \subset M$. If $f^{-1}(z)$ is an interval for each $z \in K$ we are done. Otherwise, the set Q_1 of points z in K for which $f^{-1}(z)$ is not an interval is empty. Denote the set of points for which $f^{-1}(z)$ is not an interval by Q_1. We will eliminate the points in Q_1 by a convergent sequence of modifications.

Step 1. For $z \in Q_1$, put

$$a_z = \min f^{-1}(z) \text{ and } b_z = \max f^{-1}(z).$$

Let

$$L_1 = \sup \{l_z = b_z - a_z : z \in Q_1\}.$$

Choose a sequence $\{z_k\}$ in Q_1 such that $\{l_{z_k}\}_{k=1}^{\infty}$ is increasing and converges to L_1. Choose a convergent subsequence of $\{z_k\}_{k=1}^{\infty}$ and let $z^{(1)} \in K$ be its limit. (See Definition A.6 and subsection A.4.1 in Appendix A.) Relabeling, we have $z_k \to z^{(1)}$. Some subsequence $\{a_{z_{k_j}}\}$ of the sequence $\{a_{z_k}\}$ converges to a point $a^{(1)} \in I$; the corresponding subsequence $\{b_{z_{k_j}}\}$ converges to a point $b^{(1)}$. We have $b^{(1)} - a^{(1)} = L_1$ and $f(a^{(1)}) = f(b^{(1)}) = z^{(1)}$. Put

$$f_1(z) = \begin{cases} f(z) \text{ if } z \notin [a^{(1)}, b^{(1)}] \\ z^{(1)} \text{ if } z \in [a^{(1)}, b^{(1)}] \end{cases}.$$

If $f_1(I)$ satisfies condition (3), we are done. Otherwise denote the set of points for which $f_1^{-1}(z)$ is not an interval by $Q_2 \subsetneqq Q_1$. Then repeat Step 1 for f_1. The interval $[a^{(2)}, b^{(2)}]$ is disjoint from $[a^{(1)}, b^{(1)}]$, $L_1 \geq L_2$ and

$$f_2(z) = \begin{cases} f_1(z) \text{ if } z \notin [a^{(2)}, b^{(2)}] \\ z^{(2)} \text{ if } z \in [a^{(2)}, b^{(2)}] \end{cases}.$$

If this process stops after n steps, we are done because $f_n(I)$ satisfies condition (3). Otherwise, we have a sequence $\{f_n\}$ of continuous maps with $f_{n+1}(I) \subset f_n(I)$ and $Q_{n+1} \subsetneqq Q_n$. The sequence $\{[a^{(n)}, b^{(n)}]\}$ is composed of pairwise disjoint intervals with non-increasing length $L_n \to 0$.

We have

$$f_n = f \text{ on } I - \bigcup_{j \le n} [a^{(j)}, b^{(j)}]$$

and $f_n = f(a^{(j)})$ on $[a^{(j)}, b^{(j)}]$. Consequently, if we let $\epsilon > 0$ and choose $\delta > 0$ such that

$$|t - u| < \delta \Rightarrow |f(t) - f(u)| < \epsilon,$$

then for $n \in \mathbb{N}$ we have

$$|t - u| < \delta \Rightarrow |f_n(t) - f_n(u)| < \epsilon.$$

Choose $n \in \mathbb{N}$ such that $L_n < \delta$; then for $m \in \mathbb{N}$ we have

$$\|f_{n+m} - f_n\| < \epsilon.$$

Thus, f_n converges uniformly to a continuous map g that satisfies the three conditions and $g(I) \subset \cap_{n \ge 1} f_n(I) \subset M$.

We are now ready to prove Proposition 3.2. A valid proof of this proposition entails consideration of a subtle possibility in order to avoid a sometimes fatal flaw. We follow the proof with an example to explain why the proof breaks down without the additional consideration. Then we describe a modification of the proof.

3.2.3.2 *Proof of Proposition 3.2*

If f is a one-to-one map, we are done. Otherwise, denote the set of points $z \in K = f(I)$ for which $f^{-1}(z)$ is not a single point by Q. Without loss of generality, we suppose that neither p nor q is in Q. For $z \in Q$,

$$a_z = \min f^{-1}(z) < b_z = \max f^{-1}(z)$$

and $f^{-1}(z) = [a_z, b_z]$. Because these intervals are pairwise disjoint, the sum of their lengths is less than or equal to 1 and we can put them in a list $\{[a_n, b_n]\}$, finite or countably infinite, with nonincreasing lengths converging to 0. (We shall see that it may be necessary to augment the list with some one point sets before we begin to define g.)

Using the Cantor function as a model, we will describe a process that defines a continuous nondecreasing map g from I onto I such that g is constant on each $[a_n, b_n]$ and increasing at all other points. After g is defined, we can put $h = f \circ g^{-1}$.

We begin a definition of g by putting $g(0) = 0$ and $g(1) = 1$.

Step 1. The first interval $[a_1, b_1]$ in the list does not contain 0 or 1. Define

$$g(x) = 1/2 \text{ if } x \in [a_1, b_1].$$

Compress the list by removing $[a_1, b_1]$. If the list is now empty, we extend the definition of g to all of I as follows. Define g to be linear from $(0, g(0)) = (0, 0)$ to $(a_1, g(a_1)) = (a_1, 1/2)$ on the interval $[0, a_1]$ and define g to be linear from $(b_1, g(b_1)) = (b_1, 1/2)$ to $(1, g(1)) = (1, 1)$ on the interval $[b_1, 1]$. If the list contains intervals, go to Step 2.

Step 2. If no interval in the list is between $[0]$ and $[a_1, b_1]$, define g to be linear from $(0, g(0))$ to $(a_1, g(a_1))$ on the interval $[0, a_1]$. Otherwise, choose the first interval $[a_n, b_n]$ in the list that is between $[0]$ and $[a_1, b_1]$ and define g on $[a_n, b_n]$ by

$$g(x) = (1/2)\,(g(0) + g(a_1)) = 1/4.$$

If no interval in the list is between $[a_1, b_1]$ and $[1]$, define g to be linear from $(b_1, g(b_1))$ to $(1, g(1))$ on the interval $[b_1, 1]$. Otherwise, choose the first interval $[a_m, b_m]$ in the list that is between $[0]$ and $[a_1, b_1]$ and define g on $[a_m, b_m]$ by

$$g(x) = (1/2)\,(g(b_1) + g(1)) = 3/4.$$

Compress the list by removing the intervals on which we have defined g at Step 2. If the list is now empty, define g to be linear on the segments on which g is not yet defined, and we are done. Otherwise, g is defined on a finite union of pairwise disjoint intervals, and the range of g includes the set $\{j/2^2 : 0 \leq j \leq 2^2\}$. Repeat this process.

Suppose the process does not terminate with an empty list and g is not defined on all of $[0, 1]$ at the end of any finite number of steps. After completing Step n for each positive integer n, we have g nondecreasing on its current domain; the range of g includes the dense subset

$$\{j/2^n : 0 \leq j \leq 2^n, n \in \mathbb{N}\}$$

of I. Moreover, g is defined on the first n intervals in the original list by the end of Step n, so the current domain of g includes all the intervals in the original list. Suppose (The validity of this supposition will be discussed later.) the current domain of g is dense in I. Then we define g at the points x where g is not yet defined by

$$\begin{aligned} g(x) &= \sup\{g(t) : t < x \text{ and } g(t) \text{ is defined}\} \\ &= \inf\{g(t) : x < t \text{ and } g(t) \text{ is defined}\} \end{aligned}$$

and we have g defined on I with the properties needed.

We will use g to define a one-to-one continuous map h from I onto K with $h(0) = p$ and $h(1) = q$. According to the definition of g, $g^{-1}(t)$ is either one point or some interval $[a_n, b_n]$ in the original list. However, $f^{-1}(t)$ is also either one point or some interval $[a_m, b_m]$ in the original list. Thus, we define the map h by putting

$$h(t) = f(g^{-1}(t)).$$

The following example illustrates the fact that g might not be defined on a dense subset of I after we have completed Step n for each positive integer n.

Example 3.2. Suppose the original list

$$\{[a_n, b_n]\}_{n\geq 1} = \{[1/2 + 1/2^{n+1}, 1/2 + 1/2^{n+1} + 1/4^{n+1}]\}_{n\geq 1}$$

corresponds to a map f. After all the steps are completed, the domain of g is $\{0\} \cup (1/2, 1]$ and

$$\inf\{g(t) : 1/2 < t \text{ and } g(t) \text{ is defined}\} = 0$$

because $g([a_n, b_n]) = 1/2^n$. Consequently, we must have $g([0, 1/2] = 0$, which is unacceptable.

Note that the point $1/2$ is a limit point of endpoints of intervals in the original list, but it is not an endpoint of any interval in the list. If we insert the one point set $[1/2]$ in the original list, then we define $g(1/2) > 0$ at some step and define g to be linear from $(0, g(0))$ to $(1/2, g(1/2))$ on the interval $[0, 1/2]$ at the next step. With this addition to the original list the problem in this example disappears.

We describe how to adjust the general case in the proof by modifying the original list. When $\cup_{n\geq 1}[a_n, b_n]$ is dense in I, there is no problem. (For the Cantor function, $\cup_{n\geq 1}[a_n, b_n]$ is dense in I.) When $\cup_{n\geq 1}[a_n, b_n]$ is not dense in I, denote the complement in I of the closure of $\cup_{n\geq 1}[a_n, b_n]$ by

$$G = \cup(c_k, d_k),$$

where the pairwise disjoint segments (c_k, d_k) are the components of G. If there are points that are endpoints of these segments, but are not in $\cup_{n\geq 1}[a_n, b_n]$, then we insert the one point sets corresponding to these endpoints in the initial even numbered positions in the original list before we begin to define g. Then g is defined on each interval $[c_k, d_k]$ at some step in the process, so g is densely defined when all of the steps are completed. We have completed the proof of Proposition 3.2.

In the example we have one set $[1/2]$ to insert in the list: $[1/2] = [1/2, 1/2] = [a_2, b_2]$. After this addition, in Step 2 we get

$$g([1/2]) = 1/4 = (1/2)(g(0) + g(a_1)),$$

and in Step 3 we get $g(x) = (1/2)x$ for $x \in [0, 1/2]$.

3.3 Continuous images of the Cantor set

In this section we will show that given any compact subset T of the plane, there is a corresponding map f defined on the standard Cantor set $C = C_0$ for which $f(C) = T$. We have already seen that I is a continuous image of C. We now describe a function f that maps C continuously onto I. We use base 3 for the Cantor set C and base 2 for I. Points $x = 0.x_1x_2x_3\cdots$ in C have a base 3 representation composed entirely of 0's and 2's: all of the entries x_n belong to the set $\{0, 2\}$. Let $x = 0.x_1x_2x_3\cdots \text{base } 3 \in C$ and define the map f from C to I by

$$f(x) = y = 0.\frac{x_1}{2}\frac{x_2}{2}\frac{x_3}{2}\cdots \text{base } 2. \tag{3.1}$$

Note that f maps C continuously onto I. Consequently, any curve is a continuous image of C because the composition of two continuous maps is continuous.

In the following, we approach the map f defined above in a different way; this approach will indicate how to map C onto an arbitrary compact subset T of $\mathbb{C}$. We define a sequence $\{f_n\}$ of continuous maps from C to I which converges uniformly to f. We continue to use base 3 for elements in C. For n a positive integer and any point $x^{[n]} = 0.x_1x_2x_3\cdots x_n \in C$ (each x_j is even), set

$$S_{x_1x_2x_3\cdots x_n} = [x^{[n]}, x^{[n]} + 1/3^n].$$

These sets are the 2^n intervals of length $1/3^n$ which comprise the n-th level of the construction of the Cantor set C.

Define $f_n : C \to I$, by

$$f_n(x) = 0.\frac{x_1}{2}\frac{x_2}{2}\frac{x_3}{2}\cdots\frac{x_n}{2}, x \in C \cap S_{x_1x_2x_3\cdots x_n}.$$

In particular, $f_n(x) = f_n(x^{[n]})$ for all x in $C \cap S_{x_1x_2x_3\cdots x_n}$. Thus, f_n maps C continuously onto the set of points of the form $j/2^n$ with $0 \leq j < 2^n$. Note that $\frac{j}{2^n}$ is the left endpoint of the interval $\left[\frac{j}{2^n}, \frac{j}{2^n} + \frac{1}{2^n}\right]$ in I. Because

$$\|f_{n+1} - f_n\| \leq \frac{1}{2^n},$$

the sequence $\{f_n\}$ converges uniformly to the function f defined in equation (3.1).

To show that the image of f is I, let

$$t = 0.t_1t_2t_3\cdots \in I \text{ base } 2$$

and set

$$t^{[n]} = 0.t_1t_2t_3\cdots t_n.$$

Then

$$t = \lim_{n \to \infty} t^{[n]} = \cap_n \left[t^{[n]}, t^{[n]} + \frac{1}{2^n} \right].$$

Put

$$x^{[n]} = 0.(2t_1)(2t_2)(2t_3) \cdots (2t_n) \in C.$$

Then

$$\lim_{n \to \infty} x^{[n]} = x = 0.(2t_1)(2t_2)(2t_3) \cdots = \cap_n S_{(2t_1)(2t_2)(2t_3) \cdots (2t_n)} \in C.$$

Moreover,

$$f_n(x^{[n]}) = t^{[n]}.$$

According to Theorem C.2,

$$f(x) = \lim f_n(x^{[n]}) = \lim t^{[n]} = t.$$

The above approach extends to all compact sets in $\mathbb{C}$.

Proposition 3.3. *Every compact set in $\mathbb{C}$ is a continuous image of the standard Cantor set C.*

Proof. Let T be a non-empty compact set in $\mathbb{C}$. We will define a uniformly convergent sequence $\{f_n\}$ of continuous functions from C to T. These functions f_n converge uniformly to a continuous function, $f : C \to T$, whose image $f(C) = T$.

Without loss of generality, we suppose that $T \subset I \times I$. This proof is similar to our verification that C is homeomorphic to $C \times C$. We use two levels of the construction of the Cantor set C at each stage. At Stage n, we will split each of the 4^{n-1} congruent sets of length $1/9^{n-1}$ in C into four congruent subsets of length $1/9^n$. These four subsets will correspond to the four pieces generated by splitting a square of side $1/2^{n-1}$ into four congruent squares of side $1/2^n$.

Stage 1: Define f_1

C is composed of four congruent sets

$$S_{00} = C \cap [0, 1/3^2],$$
$$S_{02} = C \cap [2/9, 1/3],$$
$$S_{20} = C \cap [2/3, 7/9],$$
$$S_{22} = C \cap [8/9, 1].$$

The set T is composed of four compact, but not necessarily disjoint, subsets

$$T_{00} = T \cap [0,1/2] \times [0,1/2],$$
$$T_{01} = T \cap [0,1/2] \times [1/2,1],$$
$$T_{10} = T \cap [1/2,1] \times [0,1/2],$$
$$T_{11} = T \cap [1/2,1] \times [1/2,1].$$

Some, but not all, of these sets may be empty.

We define f_1 below by specifying its image on each of the four sets $S_{(2j)(2k)}$. If T_{00} is not empty, then S_{00} maps onto a point t_{00} in T_{00}. If T_{00} is empty, then S_{00} maps onto a point t_{00} in the first set T_{jk} which is not empty. If T_{00} is empty, f_1 and all future iterates f_k, and eventually f, have the constant value t_{00} on S_{00}.

If T_{01} is not empty, then S_{02} maps onto a point t_{01} in T_{01}. If T_{01} is empty, then S_{02} maps onto a point t_{01} in the first set T_{jk} which is not empty. If T_{01} is empty, f_1 and all future iterates f_k, and eventually f, have the constant value t_{01} on S_{02}.

If T_{10} is not empty, then S_{20} maps onto a point t_{10} in T_{10}. If T_{10} is empty, then S_{20} maps onto a point t_{10} in the first set T_{jk} which is not empty. If T_{10} is empty, f_1 and all future iterates f_k, and eventually f, have the constant value t_{10} on S_{20}.

If T_{11} is not empty, then S_{22} maps onto a point t_{11} in T_{11}. If T_{11} is empty, then S_{22} maps onto a point t_{11} in the first set T_{jk} which is not empty. If T_{11} is empty, f_1 and all future iterates f_k, and eventually f, have the constant value t_{11} on S_{22}. Thus, we have defined $f_1 : C \to T$.

Stage 2: Define f_2

We will only consider changing the value of f_1 on sets $S_{(2j_1)(2j_2)}$ for which $T_{j_1 j_2}$ is not empty. When $T_{j_1 j_2}$ is not empty, we repeat our Stage 1 process for $S_{(2j_1)(2j_2)}$ and $T_{j_1 j_2}$. We get four subsets $S_{(2j_1)(2j_2)(2j)(2k)}$, each of length $1/3^4$ in $S_{(2j_1)(2j_2)}$. We partition the square which contains $T_{j_1 j_2}$ into four closed squares of length $1/2^2$ and denote the corresponding four intersections with T by $T_{j_1 j_2 j k}$. We do our Stage 1 process with the four subsets $S_{(2j_1)(2j_2)(2j)(2k)}$ of C and the four corresponding closed subsets $T_{j_1 j_2 j k}$ of T. After applying the process to each non-empty $T_{j_1 j_2}$, we have defined f_2.

At subsequent stages $k = 3, 4, \ldots$, corresponding functions f_k are defined. Because

$$\|f_{k+1} - f_k\| < \frac{1}{2^k},$$

the sequence $\{f_k\}$ converges uniformly to a continuous map f from C to T.

To show that f maps C onto T, let $t \in T$. We will display a point $x \in C$ for which $f(x) = t$. There is a sequence $\{j_k\}$ of 0's and 1's such that

$$t = T_{j_1 j_2} \cap T_{j_1 j_2 j_3 j_4} \cap T_{j_1 j_2 j_3 j_4 j_5 j_6} \cap \cdots .$$

The base 2 points

$$t_{j_1 j_2 \cdots j_{2n}} \in T_{j_1 j_2 \cdots j_{2n}}$$

have been defined above. Set

$$t^{[2n]} = t_{j_1 j_2 \cdots j_{2n}}.$$

Then

$$t^{[2n]} \to t.$$

For $n \geq 1$,

$$x^{[2n]} = 0.(2j_1)(2j_2)\cdots(2j_{2n}) \in S_{(2j_1)(2j_2)\cdots(2j_{2n})}.$$

Because f_n maps $S_{(2j_1)(2j_2)\cdots(2j_{2n})}$ onto $t_{j_1 j_2 \cdots j_{2n}} = t^{[2n]}$,

$$f_n(x^{[2n]}) = t^{[2n]}.$$

Observe that $\{x^{[2n]}\}$ is a convergent sequence:

$$x^{[2n]} \to x = (2j_1)(2j_2)\cdots(2j_{2n})\cdots \in C.$$

Thus,

$$f(x) = t.$$

□

Corollary 3.1. *Since a curve is a compact set in $\mathbb{C}$, every curve is a continuous image of the Cantor set.*

3.4 Subsets of $\mathbb{C}$ that are not curves

Reference [Sagan (1994)] is a good place to learn more about curves, including conditions on a compact, connected set S in $\mathbb{C}$ that are necessary and sufficient for S to be a curve.

Example 3.3. The geometric comb GC displayed in Figure 3.7 is an example of a compact, connected subset of $\mathbb{C}$ which is not a curve. The set GC is composed of the interval $I = [0, 1]$ together with a sequence of tines of length 1 pointing straight up from the points $0, 1, 1/2, 1/2^2, 1/2^3, \ldots$ in I:

$$GC = I \cup \{(0, y) : 0 \leq y \leq 1\} \cup \bigcup_{n \geq 0} \{(1/2^n, y) : 0 \leq y \leq 1\}.$$

The tine pointing up from 0 is needed to make GC a closed set. We suppose that GC is the image of a continuous map f from I to $\mathbb{C}$ and reach a contradiction as follows. Let $f(t) = (x(t), y(t))$, where x and y are continuous maps from I to I. In particular, x has the intermediate value property on I and y is uniformly continuous on I. Thus, there exists $\delta > 0$ such that if $|t - u| < \delta$, then $|f(t) - f(u)| < 1/2$. Let $\{t_n\}_{n>1}$ be a sequence of points in I with $f(t_n) = (1/2^n, 1/2)$. If $m \neq n$, there exists v between t_m and t_n for which $x(v)$ is not the base of a tine. Both $|t_m - v| \geq \delta$ and $|t_n - v| \geq \delta$ because $y(v) = 0$. Consequently, $|t_m - t_n| \geq 2\delta$ if $m \neq n$; and we have a sequence of points in I with no convergent subsequence – that is impossible. Hence, GC is not a curve.

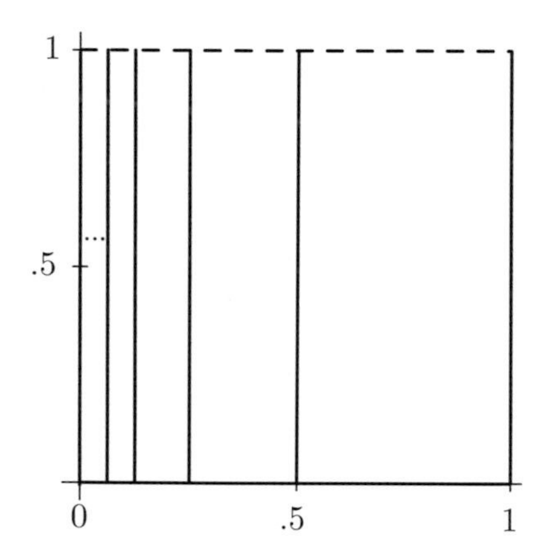

Fig. 3.7: The geometric comb. Including the dashed line gives the geometric ladder.

Example 3.4. The Cantor Comb CC has a tine of length 1 pointing straight up from each point of the Cantor set C:

$$CC = I \cup \bigcup_{x \in C} \{(x, y) : 0 \leq y \leq 1\}.$$

The Cantor Comb is not a curve. Later we consider some modifications of GC and CC which are curves.

3.5 Generalized curves

Although GC in Example 3.3 is not a curve, it is a generalized curve. Before proceeding we make an observation about images of continuous maps.

The map $F(t) = \tan(\pi t/2)$ is a one-to-one continuous map from the half open interval $[0,1)$ to the half line $[0,\infty)$; this map has a continuous inverse $G(u) = \dfrac{2}{\pi}\arctan(u)$ which maps $[0,\infty)$ onto $[0,1)$. Suppose $s : [0,1) \to \mathbb{C}$ is continuous and $s([0,1)) = S$. Then $s \circ G : [0,\infty) \to \mathbb{C}$ is continuous and $S = s \circ G([0,\infty))$. Thus, a subset S of $\mathbb{C}$ is the image of a continuous map from $[0,1)$ to $\mathbb{C}$ if and only if S is the image of a continuous map from $[0,\infty)$ to $\mathbb{C}$.

Definition 3.3. A **generalized curve** is the image of a continuous map f from the half open interval $[0,1)$ to $\mathbb{C}$.

Definition 3.4. A non-empty set S is said to be **arcwise connected** if for each pair z, w of points in S there is a curve in S that contains z and w.

A generalized curve is arcwise connected. Furthermore, because we can map $[0,\infty)$ piecewise linearly onto GC by mapping 0 to the point $(0,1)$ and proceeding as follows

$$\begin{aligned}(0,1) \to (0,0) \to (1,0) &\to (1,1) \to (1,0)\\ &\to (1/2,0) \to (1/2,1) \to (1/2,0)\\ &\to (1/4,0) \to (1/4,1) \to (1/4,0)\\ &\to \cdots\end{aligned}$$

GC is a generalized curve. The Cantor Comb in Example 3.4 is a compact, arcwise connected set that is not a generalized curve. If there were a continuous map from $[0,\infty)$ onto CC, then every tine would have to contain the image of a rational number. However, there are only countably many rational numbers, and the Cantor set is not a countable set.

We define the geometric ladder by

$$GL = GC \cup \{(x,1) : 0 \le x \le 1\}.$$

(See Example 3.3 and Figure 3.7.) The geometric ladder is also a compact, connected set that is not a curve. (See Problem 3.4.) In the next section we modify GC and GL by adding some horizontal line intervals to produce curves. Problem 3.14 introduces properties of generalized curves for you to develop.

3.6 More examples of curves

In this section we present modifications of GC and CC that are curves. We begin with an image M of the Cantor function: M is our Magic Wand. M looks like a copy of I, but, at each point $j/2^n$ of M with $0 < j/2^n < 1$, there is a continuous image of the segment $(j/3^n, (j+1)/3^n)$ in I waiting to emerge.

Consider again Figure 1.14 where the interval I lies at the base of the unit square and a copy of M lies on the top. The Cantor function maps the base to the top; the tops of the line segments from the base to the top identify the images.

Example 3.5. We modify the Cantor function. Instead of mapping the interval $[1/3, 2/3]$ to the point $(1/2, 1)$, we stretch it linearly around the (removed) triangle with vertex $(1/2, 1)$ and base segment $[1/3, 2/3]$. Similarly, we stretch the interval $[1/9, 2/9]$ linearly around the top half of the (removed) triangle with vertex $(1/4, 1)$ and base segment $[1/9, 2/9]$, where $y \geq 1/2$. We stretch the interval $[1/27, 2/27]$ linearly around the top quarter of the triangle with vertex $(1/8, 1)$ and base segment $[1/27, 2/27]$, where $y \geq 3/4$. Continuing this process, we stretch the interval $[1/3^n, 2/3^n]$ linearly around the part of the triangle with vertex $(1/2^n, 1)$ and base $[1/3^n, 2/3^n]$ where $y \geq 1 - 1/2^{n-1}$. Concurrently, we do similar stretchings to the closures of the other removed segments. The end result is a curve $W1$. Our magic wand now has triangles of depth $1/2^{n-1}$ emerging from points $((2j-1)/2^n, 1)$. Some steps in the construction are shown Figure 3.8. The verification that $W1$ is a curve is left as an exercise. (See Problem 3.6.)

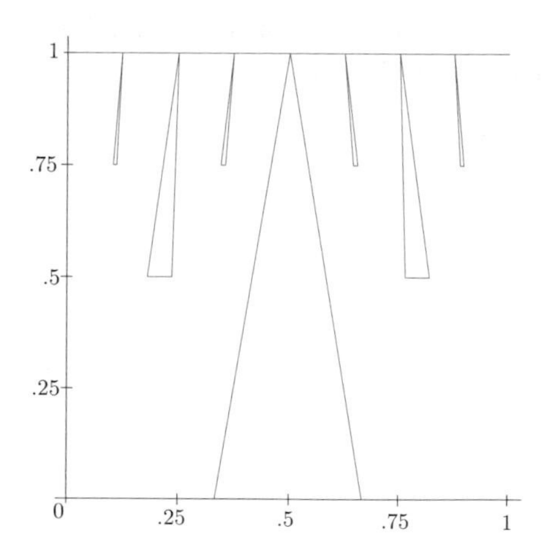

Fig. 3.8: $W1$.

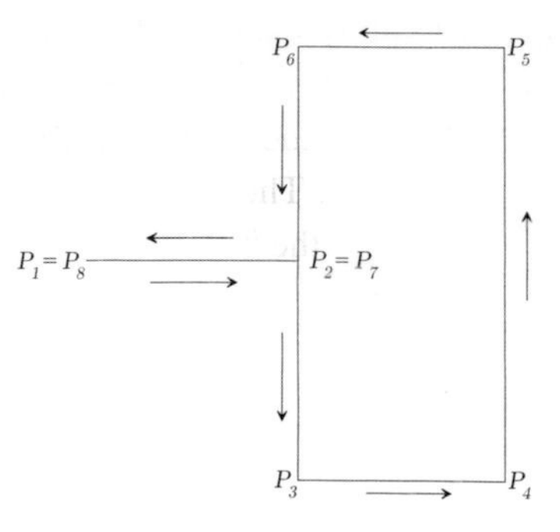

Fig. 3.9: Basic paintbrush.

Example 3.6. We begin by defining the set

$$T = \bigcup_{n \geq 1} \bigcup_{k=1}^{2^{n-1}} \left\{ \left(x, \frac{2k-1}{2^n} \right) : 0 \leq x \leq 1/2^n \right\}.$$

We place our magic wand along the y-axis:

$$M \Leftrightarrow \{(0, y) : 0 \leq y \leq 1\}.$$

Visualize our preceding example with the triangles pointing to the right. In this example, the triangles are replaced by outlines of paintbrushes (See Figures 3.9 and 3.10.) in a manner that transforms our magic wand to $GL \cup T$, which is a curve. Figure 3.11 is a graph of $GL \cup T$.

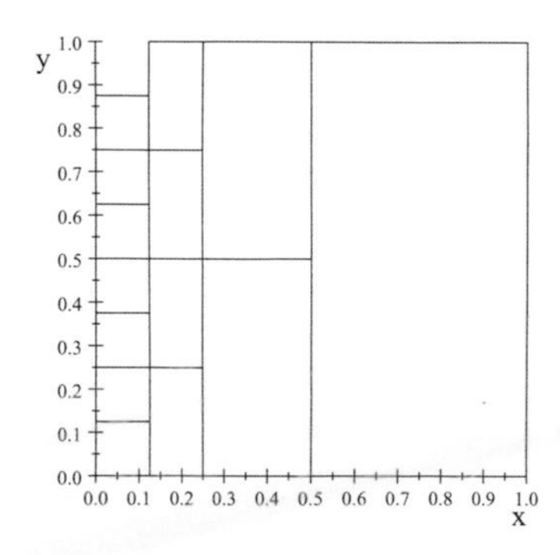

Fig. 3.10: Construction using paintbrushes.

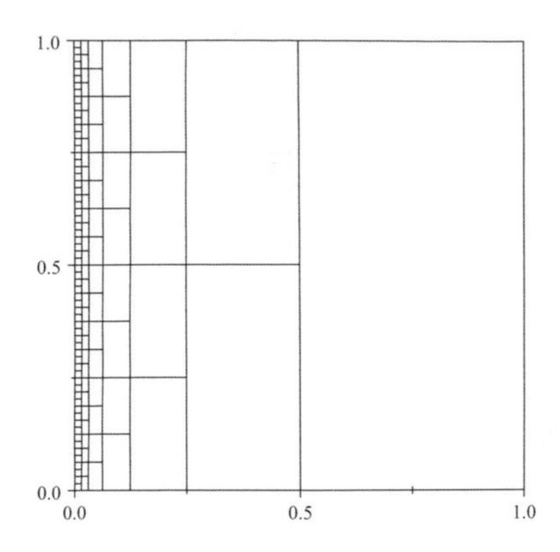

Fig. 3.11: $GL \cup T$.

Example 3.7. The circle curve shown in Figure 3.12 illustrates the flexibility of the magic wand. As introduced in Chapter 1 this curve is composed of the unit circle R and a sequence of larger concentric circles R_n of radius $1 + 1/2^n$ for $n = 0, 1, 2, 3, 4, \cdots$, together with radial spokes connecting the unit circle to the larger circles. The circles $R, R_0, \ldots, R_4$ and their spokes are displayed in Figure 3.12. To describe the function g that defines the circle curve, begin by stretching the Magic Wand uniformly around the unit circle with the ends at the point $(1, 0)$. Then set $g(0) = g(1) = (1, 0)$. From the point $g(1/4) = (0, 1)$ map the interval $(1/9, 2/9)$ onto the solid line of the construct shown in left hand side of Figure 3.13, carefully following the directional arrows. From the point $g(3/4) = (0, -1)$ map the interval $(7/9, 8/9)$ onto the reflection of the preceding construct to get the part of Stage 1 that is drawn with a dashed line. For the second stage we have

$$g\left(\frac{2k-1}{2^3}\right) = e^{\frac{(2k-1)}{4}\pi i}, \; k = 1, 2, 3, 4.$$

At each of these 2^2 points, we map the appropriate interval of length $1/3^3$ onto the construct shown in right hand side of Figure 3.13. Combining this stage to stage 1 gives the second iteration in the construction of the Circle Curve. Continue this process; when you are done, the Magic Wand has become the limit suggested by Figure 3.12.

Example 3.8. The modified Cantor comb $CC \cup V$ depicted in Figure 3.14 is a curve because, referring to Theorem 6.6, it is the connected attractor of the Iterated Function System $\{f_1, f_2, f_3, f_4, f_5\}$ composed of the five contraction

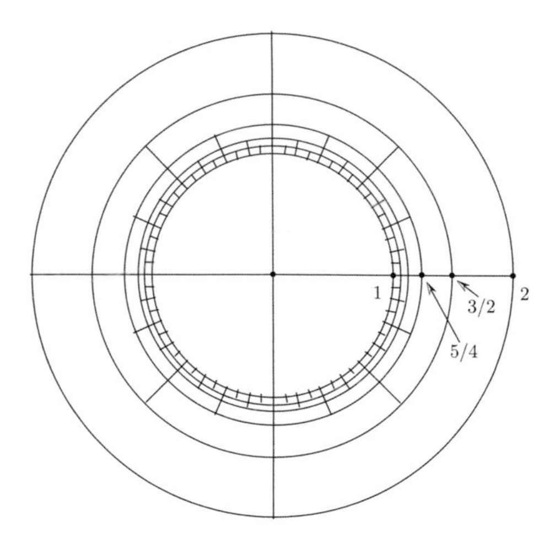

Fig. 3.12: The circle curve.

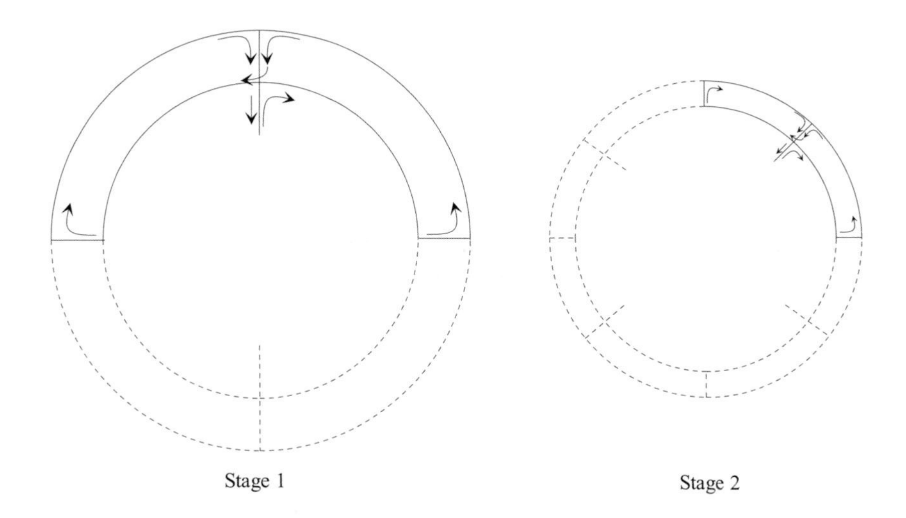

Fig. 3.13: Two stages in the construction of the circle curve.

maps displayed below.

$$\begin{aligned}
f_1(z) &= Mz, \\
f_2(z) &= (0, 1/2) + Mz, \\
f_3(z) &= (2/3, 0) + Mz, \\
f_4(z) &= (2/3, 1/2) + Mz, \\
f_5(z) &= (1/3, 0) + Nz,
\end{aligned}$$

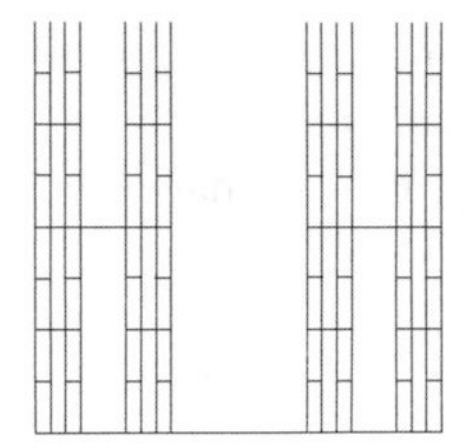

Fig. 3.14: Modified Cantor comb $CC \cup V$.

where

$$M = \begin{bmatrix} 1/3 & 0 \\ 0 & 1/2 \end{bmatrix},$$

$$N = \begin{bmatrix} 1/3 & 0 \\ 0 & 0 \end{bmatrix},$$

and $z = (x, y) = \begin{bmatrix} x \\ y \end{bmatrix}$.

The attractor is called the modified Cantor comb because its image is a modification of CC by adding a set V of horizontal intervals:

$$\begin{aligned}
V = &\{[0, 1/3] \cup [2/3, 1]\} \times \{1/2\} \\
&\qquad \text{(2 intervals of length} 1/3) \\
&\cup (\{[0, 1/9] \cup [2/9, 1/3] \cup [2/3, 7/9] \cup [8/9, 1]\} \times \{1/4, 3/4\}) \\
&\qquad \text{(8 intervals of length } 1/9) \\
&\cup (\{[0, 1/27] \cup [2/27, 1/9] \cup [2/9, 7/27] \\
&\quad \cup [8/27, 1/3] \cup [2/3, 19/27] \cup [20/27, 7/9] \cup [22/27, 23/27] \\
&\quad \cup [8/9, 25/27] \cup [26/27, 1]\} \times \{1/8, 3/8, 5/8, 7/8\}) \\
&\qquad \text{(32 intervals of length } 1/27) \\
&\cup \cdots
\end{aligned}$$

To verify that the attractor $CC \cup V$ is connected, begin the deterministic algorithm with the unit square, focus on the points $(0, 1/2)$, $(1/3, 0)$, $(2/3, 0)$, and $(2/3, 1/2)$, and observe that each iterate is connected.

Example 3.9. This example shows that the connected intersection of two curves need not be a curve or even a generalized curve. We begin by modifying $GL \cup T$ and $CC \cup V$ so that the horizontal intervals are located at irrational heights. Choose an irrational number w slightly smaller than 1, say $w = 1 - \sqrt{2}/100$.

Define a homeomorphism h of U by specifying that for $0 \leq y \leq 1/2$, $(x, y) \rightarrow (x, wy)$, and for $1/2 \leq y \leq 1$, $(x, y) \rightarrow (x, (2-w)y) + (w-1))$. The horizontal intervals in $h(GL \cup T)$ and $h\,(CC \cup V)$ have irrational heights, so

a. $h(GC \cup T)$ is a curve and

$$(GC \cup T) \cap h(GC \cup T) = GC$$

is not a curve, but it is a generalized curve.

b. $h(CC \cup V)$ is a curve and

$$(CC \cup V) \cap h(CC \cup V) = CC$$

is not a generalized curve.

In Problem 3.7 you will verify that $h(GC \cup T)$ and $h(CC \cup V)$ are curves.

Example 3.10. A fat Cantor comb C_hC is defined by replacing the standard Cantor set C by a Cantor set C_h of positive length. Then C_hC is modified to a curve $C_hC \cup V$ which has positive area at a point (x, y) if $x \in C_h$ and $y \in I$. You can show that (see Problem 3.8)

$$\begin{aligned}\operatorname{area}(C_hC \cup V) &= \operatorname{area}(C_hC) = \operatorname{len}(C_h)\\ &= \operatorname{area}((x, y) : x \in C_h \text{ and } y \in I).\end{aligned}$$

Our next definition, local connectedness, expresses a property that GC and CC gain when they are modified.

Definition 3.5. A set X in $\mathbb{R}^n$ is **locally connected** if for every point $p \in X$ and every $\epsilon > 0$ there is an $\eta(p, \epsilon) > 0$ such that for every point $q \in B(p, \eta(p, \epsilon)) \cap X$ there is a compact and connected subset of X that contains p and q and is contained in $B(p, \epsilon)$.

Problem 3.9 asks you to verify that $GC \cup T$ and $CC \cup V$ are locally connected. The Hahn-Mazurkiewicz Theorem [Sagan (1994)] asserts that a compact, connected set S in $\mathbb{R}^2$ is a curve if, and only if, S locally connected. Problem 3.9 and the Hahn-Mazurkiewicz Theorem determine that $GC \cup T$ and $CC \cup V$ are curves. We have exhibited explicit maps for these two curves. A beautiful application of the Hahn-Mazurkiewicz Theorem appears in Chapter 6.

3.7 Problems

PROBLEM 3.1. Show that the area of the Koch curve is equal to 0.

PROBLEM 3.2. Complete the following.

(a) Determine the points in U that are points of density 1.
(b) Show that not every point of a curve can be a point of density 1.
(c) Construct a simple curve with a point of density 1. (One option is to simplify the construction in Example 3.1 to yield a simple curve with density one at the point $(1/2, 1/2)$.)
(d) Construct a simple curve with a dense set of points of density 1.
(e) Show that no simple curve S can contain a non-trivial curve K such that each point of K is a point of density 1 of S.

PROBLEM 3.3. Let

$$S = I \cup \{(0, y) : 0 \leq y \leq 1\} \cup \bigcup_{n \geq 1} \{(1/2^n, y) : 0 \leq y \leq 1/n\}.$$

Show that S is a curve.

PROBLEM 3.4. Show that GL is not a curve.

PROBLEM 3.5. Suppose we modify CC by shortening the tines: For each $x \in C$, choose $a_x > 0$ and put

$$A = I \cup \{(0, y) : 0 \leq y \leq 1\} \cup \bigcup_{x \in C} \{(x, y) : 0 \leq y \leq a_x\}.$$

Note that A is a Cantor comb with tines of arbitrary length. Show that A is not a generalized curve.

PROBLEM 3.6. Verify that $W1$ introduced in Example 3.5 is a curve.

PROBLEM 3.7. Explain why $h(GC \cup T)$ and $h(CC \cup V)$ are curves.

PROBLEM 3.8. Explain why $C_h C \cup V_h$ is a curve and has area equal to the length of C_h. (The set V_h must be appropriately defined.)

PROBLEM 3.9. Show that $GC \cup T$ and $CC \cup V$ are compact, connected, locally connected.

PROBLEM 3.10. Show that a compact subset S of $\mathbb{C}$ is connected if for each pair $\{a, b\}$ of points in S and each $\epsilon > 0$, there exists a finite sequence $\{t_j\}_{0 \leq j \leq n}$ of points in S such that $t_0 = a$, $t_n = b$, and $|t_j - t_{j-1}| \leq \epsilon$ for $1 \leq j \leq n$.

PROBLEM 3.11. Suppose that E is a non-trivial curve in $\mathbb{C}$. Show that there is a nowhere locally constant, continuous map m from I to $\mathbb{C}$ for which $m(I) = E$.

Table 3.4: Unknown matrices.

j	0	1	2	3
$C(j)$	$(0,0)$	$(1/2,0)$	(a,b)	$(1/2,0)$
$V(j)$	$1/2$	$M(1)$	$M(2)$	$1/2$

PROBLEM 3.12. Let K be a compact set in $\mathbb{R}^2$. Let m map the Cantor set C continuously onto K. Let m be linearly extended to I. Show that the $\text{area}\,(m\,(I)) = \text{area}\,(K)$.

PROBLEM 3.13. We use base 4 and the values in the table below for this problem. We wish to show that a generic triangle, including the interior, is a curve by exhibiting a continuous map f from I onto the triangle. Without loss of generality, we suppose that a longest side of the triangle is the interval $[0,1]$ and the third vertex is at the point (a,b), where $a > 0$ and $a^2 + b^2 \leq 1$. For $x = 0.x_1x_2\cdots \in I = [0,1]$ set $f_0(x) = (x,0)$.

(a) Referencing Figure 3.15 which shows the orientation of the subtriangles, find the two 2×2 matrices $M\,(1)$ and $M\,(2)$ in Table 3.4 so that the sequence of functions $\{f_n\}$ given by

$$f_{n+1}(x) = C(x_1) + V(x_1)f_n(\ell x).$$

converges to the desired function f.

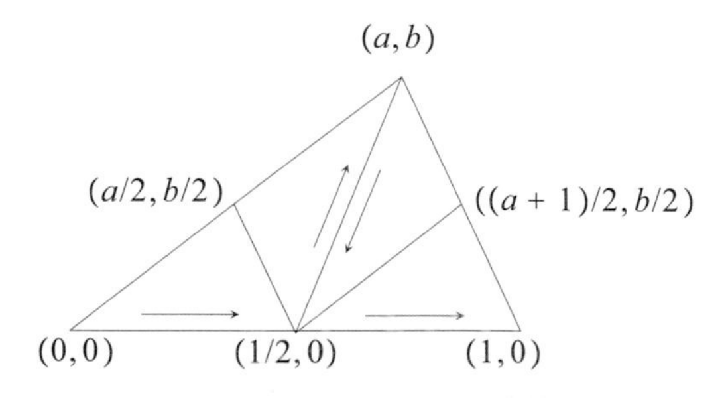

Fig. 3.15: Orientation of subtriangles.

(b) Draw $f_1(I)$, $f_2(I)$, and $f_3((I)$.

(c) Show that the sequence $\{f_n\}$ converges uniformly to a continuous map f whose image is the original triangle.

PROBLEM 3.14. **Project.** Write a paper containing examples of generalized curves in $\mathbb{C}$ noting the following:

(a) Generalized curves need not be closed and they need not be bounded; for example, $\mathbb{C}$ is a generalized curve.
(b) A curve is a generalized curve.
(c) If a finite union of curves is a connected set, then it is a curve.
(d) A generalized curve is arcwise connected.
(e) An arcwise connected set is connected.
(f) Suppose that

$$H = \{(0, y) : -1 \leq y \leq 1\} \cup \{(x, \sin(1/x) : 0 < x \leq 1\}.$$

Show that G is the compact, connected union of a curve and a generalized curve but H is not arcwise connected.
(g) Show that a subset S of $\mathbb{R}^2$ is a generalized curve if and only if S is an arcwise connected countable union of curves.
(h) Explain why an arcwise connected, countable union of generalized curves is a generalized curve.

PROBLEM 3.15. Display some of the interesting curves that you can construct with Magic Wands.

PROBLEM 3.16. Show that the curve $g_h(I)$ has a tangent line at a point $g_h(p)$ if and only if $p \notin C_h$.

Chapter 4

Generalizations of the Koch curve

Generalizations of the Koch curve start with a generator $G(a, \theta)$ as shown in Figure 4.1. Results of our investigations of the two-parameter family of curves generated by $G(a, \theta)$ appear in [Darst *et al.* (2008)]. Note that the Koch curve is generated by starting with $G(1/3, \pi/3)$. Curves generated by $G(a, \pi/3)$ with $a \neq 1/3$ appear in Plate 56 of [Mandelbrot (1983)] and Example 9.5 in [Falconer (1990)]. We learned from Keleti's work [Keleti (2006)] that in 1998 M. van den Berg asked the following question at a colloquium talk at University College London: For which values of a does $G(a, \pi/3)$ generate a simple curve? Keleti gave the answer: $G(a, \pi/3)$ generates a simple curve when $a > 1/4$.

In this chapter we present properties of the curve generated by $G(1/4, \pi/3)$ and the corresponding curve for $\theta = \pi/4$ generated by $G(1-\sqrt{3}/2, \pi/4)$. Neither curve is a simple curve; each curve contains points that are images of two points in I (double image points: double points). We exploit the self-similarity in the construction of the curves to characterize the double points. Triples of double points in the first curve compose the vertices of equilateral triangles. A new class of Cantor sets is used to describe the double points in the second curve.

4.1 Construction of generalizations

We begin our discussion of generalizations of the Koch curve by considering the generic generator $G = G(a, \theta)$ as shown in Figure 4.1, allowing the angle θ and the length a to vary. If $L < 1$ then G generates a self-similar curve by a replacement process similar to that for the Koch curve presented in Section 2.1. For $\theta = \pi/3$ examples of a few stages in the process for various values of a are shown in Table 4.2.

For this discussion we consider the generator G as a path in the complex plane from $(0, 0)$ to $(1, 0)$. We will make use of the constants $C(j)$ and $D(j)$ given in

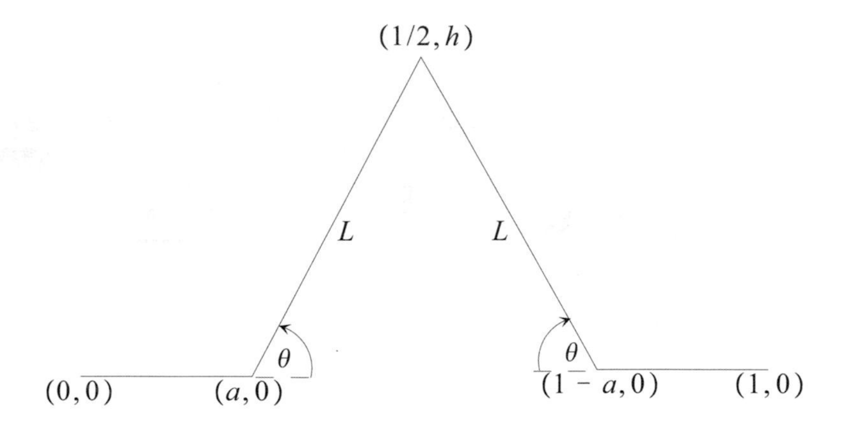

Fig. 4.1: Generic generator.

Table 4.1: Values of $C(j)$ and $D(j)$.

j	0	1	2	3	4
$C(j)$	0	a	$(1/2, h)$	$1-a$	1
$D(j)$	a	$Le^{i\theta}$	$Le^{-i\theta}$	a	$-$

Table 4.1 to describe this path. Throughout, we restrict attention to the setting $0 < a < 1/2$ and $0 < L < 1$. The path is composed of four intervals that abut in pairs at three interior nodes $C(1)$, $C(2)$, and $C(3)$. In some sense, G is a classic generator of fractal images. For example, when $a = 1/3$ and $\theta = \pi/3$, we are in the setting of the Koch curve, which is a simple curve. If $a = 0$ and $\theta = \pi/4$, you generate the Levy curve [Edgar (1993)], which is not a simple curve because it is known to have an interior [Bailey *et al.* (2002)].

With $\theta = \pi/3$ we will see that a simple curve is generated by $G(a, \theta)$ if and only if $a > 1/4$, and we will examine the curve generated by $G(1/4, \pi/3)$. With $\theta = \pi/4$ we will show that $G(a, \theta)$ is a simple curve when $a > 1 - \sqrt{3}/2$, and we will examine the curve generated by $G(1 - \sqrt{3}/2, \pi/4)$. The set of points (a, θ) in $\mathbb{R}^2$ for which $G(a, \theta)$ generates a simple curve is an open set in $\mathbb{R}^2$. We are focusing on the two boundary points of this set that correspond to $\theta = \pi/3$ and $\theta = \pi/4$.

We begin by defining a natural homeomorphism f_1 from the interval $I = [0, 1]$ onto G. The mapping f_1 is specified by requiring that it map each interval

Table 4.2: Iterations with angle of $\pi/3$ using various values of a. The graph for $k = 1$ is not included since it reduces to a special case of the generator.

a	$k = 2$	$k = 3$	Limit curve
$\frac{1}{3}$			
$\frac{7}{25}$			
$\frac{1}{4}$			
$\frac{1}{5}$			

$[j/4, (j+1)/4]$ in I linearly onto the corresponding interval $[C(j), C(j+1)]$ in G, $j = 0, 1, 2, 3$. Beginning with f_1 we recursively define a sequence of continuous maps $\{f_k\}$ on I which converges uniformly to a continuous map f from I into $\mathbb{C}$. The set $K_{a,\theta}$ is defined to be the image $f(I) = \{f(x) : x \in I\}$. The steps in the creation of the Koch curve given in Chapter 2 (the case $a = 1/3$, $\theta = \pi/3$) demonstrate the process.

4.1.1 *The iteration process*

Because the generator is composed of four intervals, we use base 4 representations for numbers $x = 0.x_1x_2x_3\cdots$ base 4 in I. Define

$$D(j) = C(j+1) - C(j), j = 0, 1, 2, 3, \tag{4.1}$$

$$f_1(x) = C(x_1) + D(x_1)\ell x$$

and

$$f_{k+1}(x) = C(x_1) + D(x_1)f_k(\ell x) \text{ for } k \geq 1. \tag{4.2}$$

Then

$$\|f_{k+1} - f_k\| = D\,\|f_k - f_{k-1}\|$$

where

$$D = \max\{|D(k)| : k \in \{0, 1, 2, 3\}\} = \max\{a, L\} < 1.$$

Consequently, $\{f_k\}$ is a Cauchy sequence of continuous functions and must converge to a continuous function on I. (See Theorem C.6 and Corollary C.2 in Appendix C.) Moreover, from formula (4.2), f satisfies the following functional equation

$$f(x) = C(x_1) + D(x_1)f(\ell x), x \in I. \tag{4.3}$$

We call the graph $K_{a,\theta}$ of f a **generalized Koch curve**.

Notice that in general the function f depends on a and θ. If we need to emphasize this dependence, we will do so by writing, for example, f_a or $f_{a,\theta}$.

4.1.2 *A decomposition of $K_{a,\theta}$*

The set $K_{a,\theta}$ is composed of four *parts* $K_{a,\theta} = T_0 \cup T_1 \cup T_2 \cup T_3$, where

$$T_j = f\left(\left[0.j, 0.j + \frac{1}{4}\right]\right) = C(j) + D(j)f(I), j = 0, 1, 2, 3.$$

(T_0 is depicted for a particular curve by the shaded region in Figure 4.3.) Each T_j is composed of four parts $T_j = T_{j0} \cup T_{j1} \cup T_{j2} \cup T_{j3}$,

$$T_{jk} = f\left(\left[0.jk, 0.jk + \frac{1}{4^2}\right]\right), k = 0, 1, 2, 3.$$

Inductively, each $T_{j_1 j_2 \cdots j_n}$ is composed of the four parts

$$T_{j_1 j_2 \cdots j_n} = \bigcup_{j_{n+1}=0}^{3} T_{j_1 j_2 \cdots j_n j_{n+1}},$$

where

$$T_{j_1 \cdots j_{n+1}} = f\left(\left[0.j_1 j_2 \cdots j_n j_{n+1}, 0.j_1 j_2 \cdots j_n j_{n+1} + \frac{1}{4^{n+1}}\right]\right) \tag{4.4}$$

and $j_{n+1} = 0, 1, 2, 3$. Moreover, for each $x = 0.x_1 x_2 x_3 \cdots$ base $4 \in I$,

$$f(0.x_1 x_2 x_3 \cdots) = \bigcap_{n \geq 1} T_{x_1 x_2 x_3 \cdots x_n} = T_{x_1 x_2 x_3 \cdots}.$$

The functional equation for f given in equation (4.3) implies that each part is a suitably reduced, translated, and rotated copy of $K_{a,\theta}$. In particular, T_0 is an a-sized copy of $K_{a,\theta}$; T_1 is a L-sized copy of $K_{a,\theta}$, rotated counterclockwise by an angle of θ about $(0, 0)$ and then translated so that $(0, 0)$ maps to $C(1)$; T_2 is a L-sized copy of $K_{a,\theta}$ rotated by an angle of $\pi/3$ about $(0, 0)$ clockwise and then translated so that $(0, 0)$ maps to $C(2)$; T_3 is an a-sized copy of $K_{a,\theta}$ translated to $C(3)$.

4.2 Double points in $K_{a,\theta}$ with $\theta = \pi/3$ and $a = 1/4$

Throughout this section θ is fixed at $\pi/3$. Experimentally, computer pictures of the approximation curves $P^j_{a,\pi/3}$ indicate that the map f is a homeomorphism from I to $K_{a,\pi/3}$ if $1/4 < a < 1/2$ and is not a homeomorphism if $0 < a < 1/4$. Keleti [Keleti (2006)] has shown that although all the f_k's are homeomorphisms, when $a = 1/4$, f is not. So $a = 1/4$ is the pivotal value.

Now fix $a = 1/4$ so that $L = 1 - 2a = 1/2$ and $h = \sqrt{3}\,(1/2 - a) = \sqrt{3}/4$. We use the drawing in Figure 4.2 as an illustrative approximation for $K_{1/4,\pi/3}$. We show that the point w indicated on the expanded display of RS given in Figure 4.3 is a double point of RS – the fractal $K_{1/4,\pi/3}$ is named RS for the canine companions of the first author, Rosie and Sparky, since it looks like two dogs joined by a leash; w is the point where the leash is attached to Rosie.

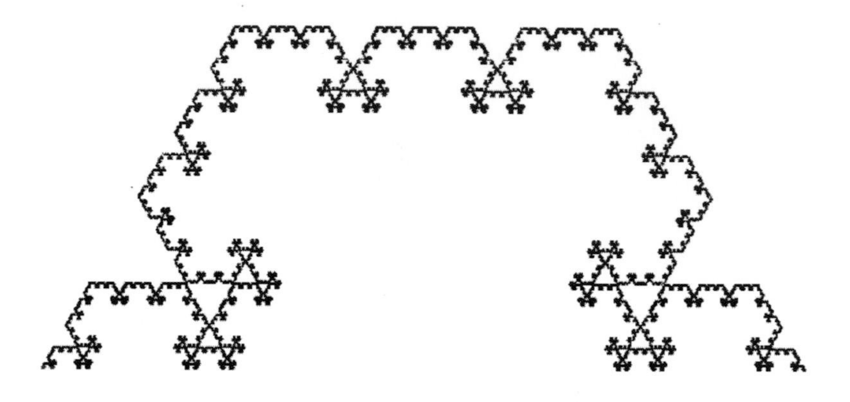

Fig. 4.2: Rosie and Sparky with leash.

First, we will argue geometrically that $f\,(3/20) = f\,(27/80) = w$. Then we will use the functional equation (4.3) for f to show that $w = \frac{3}{16} + \frac{\sqrt{3}}{16}i$. From the discussion in subsection 4.1.2, RS $= T_0 \cup T_1 \cup T_2 \cup T_3$, where

$$T_0 = \frac{1}{4}\text{RS},$$

$$T_1 = \frac{1}{4} + \frac{1}{2}e^{i\pi/3}\text{RS},$$

$$T_2 = \left(\frac{1}{4} + \frac{1}{2}e^{i\pi/3}\right) + \frac{1}{2}e^{-i\pi/3}\text{RS},$$

$$T_3 = \frac{3}{4} + \frac{1}{4}\text{RS}.$$

In a careful examination of Figure 4.3, w appears to be in T_0 and in T_1. We need to find the point P_1 in $[0, 1/4]$ and the point P_6 in $[1/4, 1/2]$ that map onto w. (The choice of the subscripts in labeling these points will become apparent shortly.) To find P_1, we focus on Figure 4.3 which contains an enlarged copy of T_0, together with spin-off copies of some of its parts. Notice that w appears to be in T_{02}, T_{021}, and in T_{0212}. The repeating pattern is clear: $P_1 = 0.0\underline{21}\,\text{base}\,4 = 3/5$. Similarly, starting with T_1, we get $P_6 = .11\underline{12}\,\text{base}\,4 = 27/80$. The simple forms of the base 4 representations of P_1 and P_6 make it easy to find their function

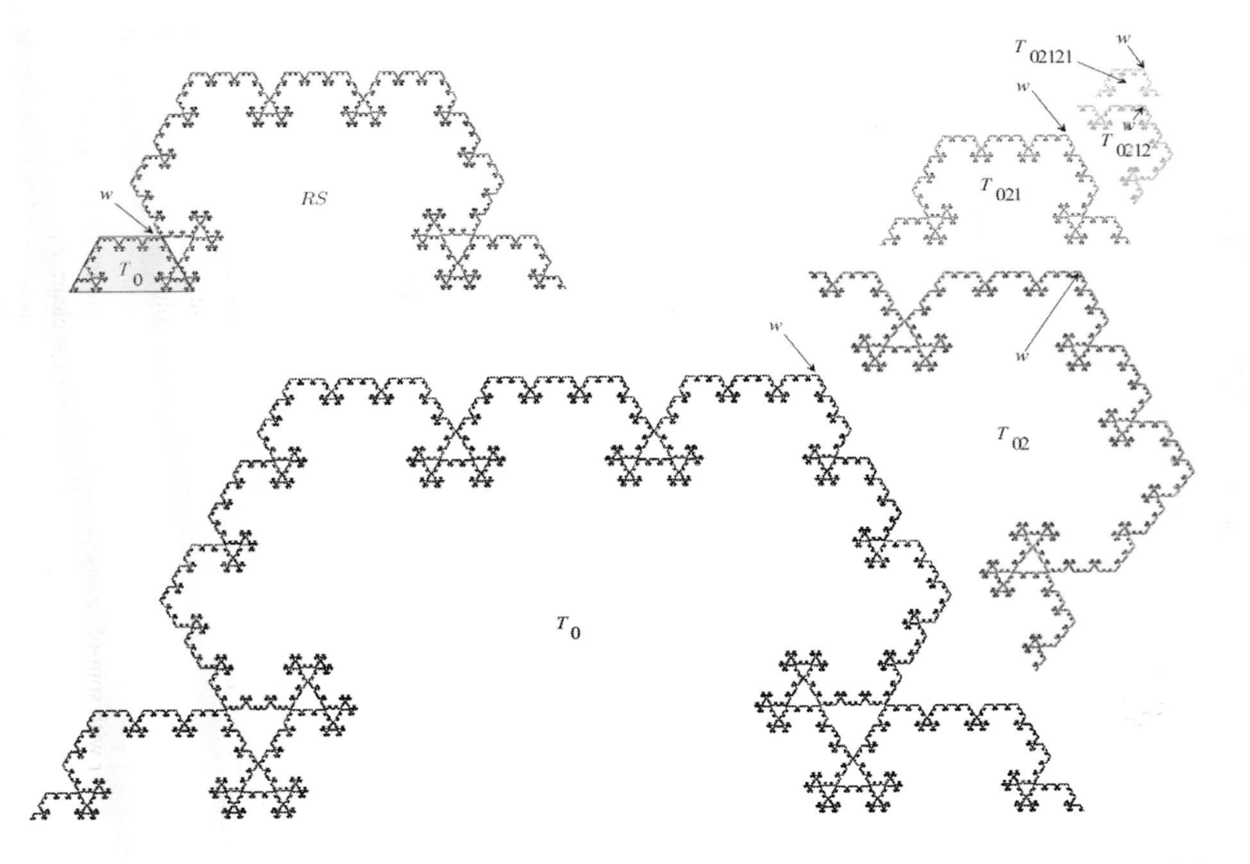

Fig. 4.3: Parts of RS used to determine the double point w. The shaded area covers the T_0 part of RS.

values. First, observe that

$$\begin{aligned} f(.\underline{21}) &= C(2) + D(2)f(0.1\underline{21}) \\ &= C(2) + D(2)(C(1) + D(1)f(0.\underline{21})) \\ &= C(2) + D(2)C(1) + D(2)D(1)f(0.\underline{21}). \end{aligned}$$

We solve this last equation for $f\,(0.\underline{21})$ and obtain

$$f(.\underline{21}) = \frac{C(2) + D(2)C(1)}{1 - D(2)D(1)} = \frac{3}{4} + \frac{\sqrt{3}}{4}i. \tag{4.5}$$

Hence,

$$f(P_1) = f(0.0\underline{21}) = \frac{1}{4}f(0.\underline{21}) = \frac{3}{16} + \frac{\sqrt{3}}{16}i.$$

Now compute

$$\begin{aligned} f(P_6) &= f(0.111\underline{21}) \\ &= C(1) + D(1)\left(C(1) + D(1)\left(C(1) + D(1)f(.\underline{21})\right)\right) \\ &= \frac{1}{4} + \left(\frac{1}{2}e^{\frac{i\pi}{3}}\left(\frac{1}{4} + \frac{1}{2}e^{\frac{i\pi}{3}}\left(\frac{1}{4} + \frac{1}{2}e^{\frac{i\pi}{3}}\left(\frac{3}{4} + \frac{\sqrt{3}}{4}i\right)\right)\right)\right) \\ &= \frac{3}{16} + \frac{\sqrt{3}}{16}i. \end{aligned}$$

Thus, we have located one double point $w = \frac{3}{16} + \frac{\sqrt{3}}{16}i = f(P_1) = f(P_6)$. In fact, the six points

$$P_1 = 0.0\underline{21}\text{ base }4 < P_2 = 0.02\underline{21}\text{ base }4 < P_3 = 0.101\underline{21}\text{ base }4$$

$$< P_4 = 0.10\underline{21}\text{ base }4 < P_5 = 0.1111\underline{21}\text{ base }4 < P_6 = 0.111\underline{21}\text{ base }4$$

map onto three double points $w, u = \dfrac{7}{32} + \dfrac{\sqrt{3}}{32}i$, u, $v = \dfrac{1}{4} + \dfrac{\sqrt{3}}{16}i$, v and w, respectively. The three double points u, v and w comprise the corners of an equilateral triangle which is easy to spot in Rosie.

The angle $\pi/3$ is special in the sense that both w and u are double points. Looking again at Figure 4.3, we see that T_{02} and T_{10} are symmetric with respect to the line $x = a\,(1 - a/2) = 7/32$. The point u is on this line; T_{02} and T_{10} touch only at u. We will use this symmetry to study a second variation of the Koch curve in Section 4.3.

In general, by symmetry, if $f(x)$ is a double point, then $f(1-x)$ is a double point. For example, using base 4

$$1 - 0.0\underline{21} = 0.3\underline{12} = 0.31\underline{21} \neq 1 - 0.111\underline{21} = 0.222\underline{12} = 0.2221\underline{21},$$

and $f(0.3\underline{12}) = f(0.222\underline{12})$ is the point in Sparky that corresponds to w in Rosie. Also, if x and y map to the same point so that $f(0.x_1 \cdots) = f(0.y_1 \cdots)$ is a double point, then for any $t_1, t_2, \cdots, t_k, 0 \leq t_j \leq 3$, the points $0.t_1 t_2 \cdots t_k x_1 \cdots$ and $0.t_1 t_2 \cdots t_k y_1 \cdots$ are also double points. So we conclude that RS has infinitely many double points.

Referring to Figure 4.2, view RS as a complex atoll in $\mathbb{C}$; RS surrounds an infinity of equilateral, triangular, fractal lagoons. Lagoons are the bounded components of the complement of RS in $\mathbb{C}$. Each of the three corners of a lagoon is a double point (the image of two points in I) of the map *f*, and the inverse image of each corner is composed of two rational points in I. The image in Figure 4.4 corresponds to an enlargement of part of RS, displaying several of the equilateral triangular shaped regions.

Fig. 4.4: Enlarged view of a "triangle" of RS.

4.2.1 *The pivotal value* $a(\theta) = 1/4$ *for* $\theta = \pi/3$

For arbitrary $f_{a,\theta}$ the pivotal value $a(\theta)$ of a corresponding to θ has the property that $K_{a(\theta),\theta}$ has double points but the map $f_{a,\theta}$ is a homeomorphism if $a(\theta) < a < 1/2$. How do we find $a(\theta)$?

Note that T_{02} and T_{10} are symmetric with respect to the vertical line $x = a\,(1 - a/2)$. If $a > 1/4$, neither of the two pieces touches the line, and if $a < 1/4$, they both cross the line. Focusing on T_{02}, we see that $1/4$ is the value of a for which

$$\max\{x : (x, y) \in T_{02} \text{ of } K_a\} = a\left(1 - \frac{a}{2}\right).$$

Since T_0^a is an a-sized copy of K_a, we can find $a\,(\pi/3)$ by solving the equation

$$\max\{x : (x, y) \in T_2 \text{ of } K_a\} = 1 - \frac{a}{2}.$$

Recall that $L = 1 - 2a$ and refer to Figure 4.5 which shows (following the dashed lines) that

$$\begin{aligned}\max\{x : (x, y) \in T_2 \text{ of } K_a\} &= a + \frac{1}{2}\frac{L}{2} + \frac{L}{2} + \frac{L^2}{2} + \cdots \\ &= a + \frac{1-2a}{4} + \frac{L}{2}\left(\frac{1}{1-L}\right) \\ &= \frac{2a+1}{4} + \frac{1-2a}{2}\left(\frac{1}{2a}\right).\end{aligned}$$

Putting the preceding expression equal to $1 - a/2$ and solving for a, we find that $a = 1/4$.

When $\theta < \pi/3$ the preceding discussion applies with appropriate modification. The pivotal value $a(\theta)$ corresponding to θ can be determined by solving the equation

$$\max\{x : (x, y) \in T_2 \text{ of } K_{a,\theta}\} = 1 - \frac{a}{2}$$

as we shall see later.

4.3 Investigation of $K_{a,\pi/4}$

Now consider the case where $\theta = \pi/4$ and therefore $0 < L = \frac{\sqrt{2}}{2}\,(1 - 2a) < 1$. We will demonstrate that the pivotal value in this case is $a = a\,(\theta) = 1 - \sqrt{3}/2$, a number not exposed by computer sketches. In this variation $K_{a,\pi/4}$ can also be viewed as a complex atoll which surrounds an infinity of fractal lagoons. However, $K_{a,\pi/4}$ has Cantor sets of double points and lagoons with various shapes.

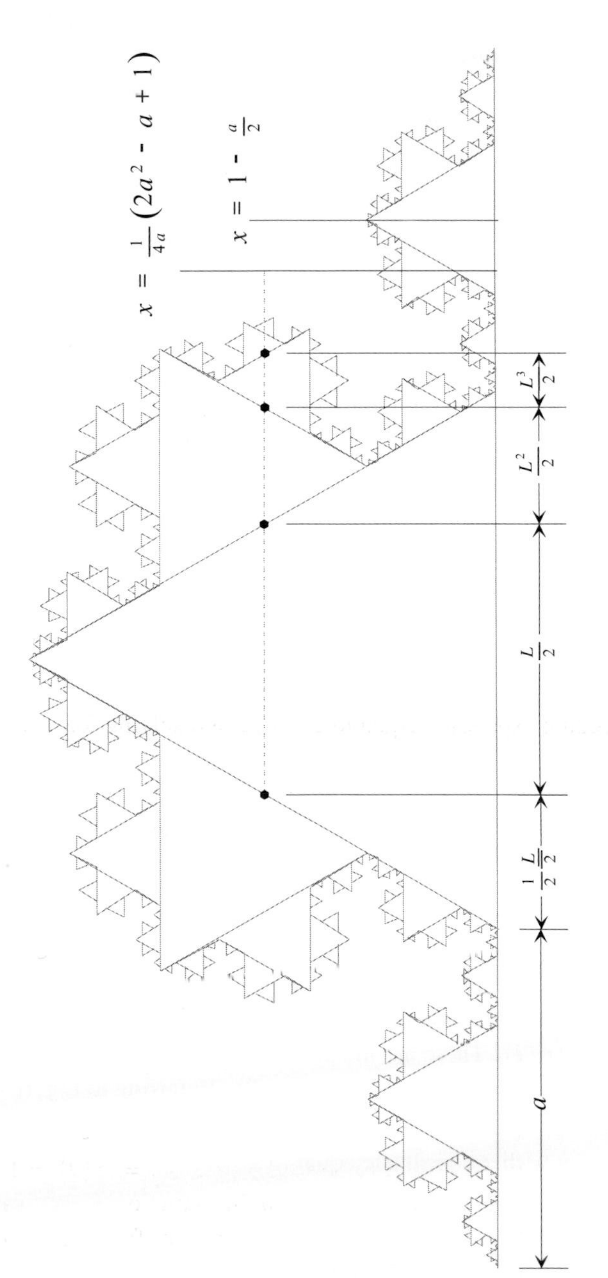

Fig. 4.5: Finding the maximum distance for $\theta = \pi/3$.

Decreasing the angle with a fixed causes T_1 and T_2 to become smaller. With the angle less than $\pi/3$, there are no double points corresponding to the points v and w in RS; thus we concentrate on double points corresponding to u. An example of $K_{a,\pi/4}$ is shown in Figure 4.6. An enlarged section is shown in Figure 4.8 and shows a Cantor set of double points beginning to form on a vertical line through the center of the image.

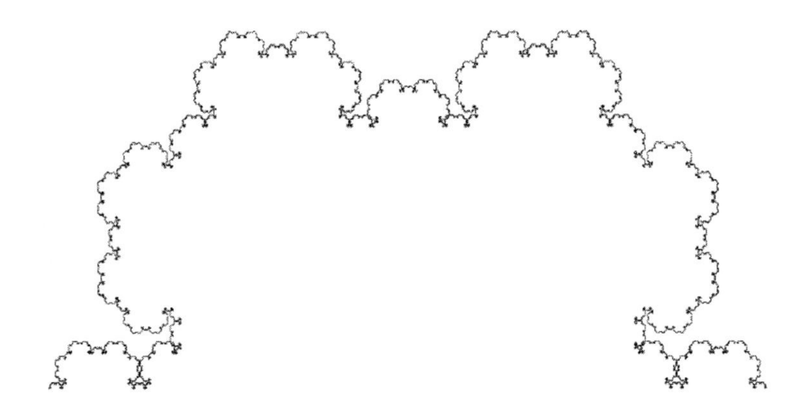

Fig. 4.6: Generalized Koch curve with $a = 1 - \sqrt{3}/2$ and $\theta = \pi/4$.

4.3.1 *Verifying* $a(\theta) = 1 - \frac{\sqrt{3}}{2}$ *for* $\theta = \pi/4$

We verify that $a(\pi/4) = 1 - \sqrt{3}/2$ is the pivotal value of a corresponding to $\theta = \pi/4$. Focusing on Figure 4.7, we see that the height H_a of $K_{a,\pi/4}$ satisfies the equation

$$H_a = \frac{\sqrt{2}}{2}(1-a)L(1 + L^2 + L^4 + \cdots) = \frac{\sqrt{2}}{2}(1-a)L\frac{1}{1-L^2}. \quad (4.6)$$

The double points corresponding to u in RS comprise the intersection of T_{022} and the line $x = a(1 - a/2)$. These are the points corresponding (via sizing by a) to the intersection of T_{22} and the line $x = (1 - a/2)$. Referring to Figure 4.6 again, when $\theta = \pi/4$, T_{22} is an L^2-sized copy of $K_{a,\pi/4}$, rotated clockwise $90°$ and translated to a point with x coordinate equal to $\frac{1}{2} + \frac{\sqrt{2}}{2}(1-a)L$. Thus, we need to solve the equation

$$1 - \frac{a}{2} = \frac{1}{2} + \frac{\sqrt{2}}{2}(1-a)L + L^2 H_a.$$

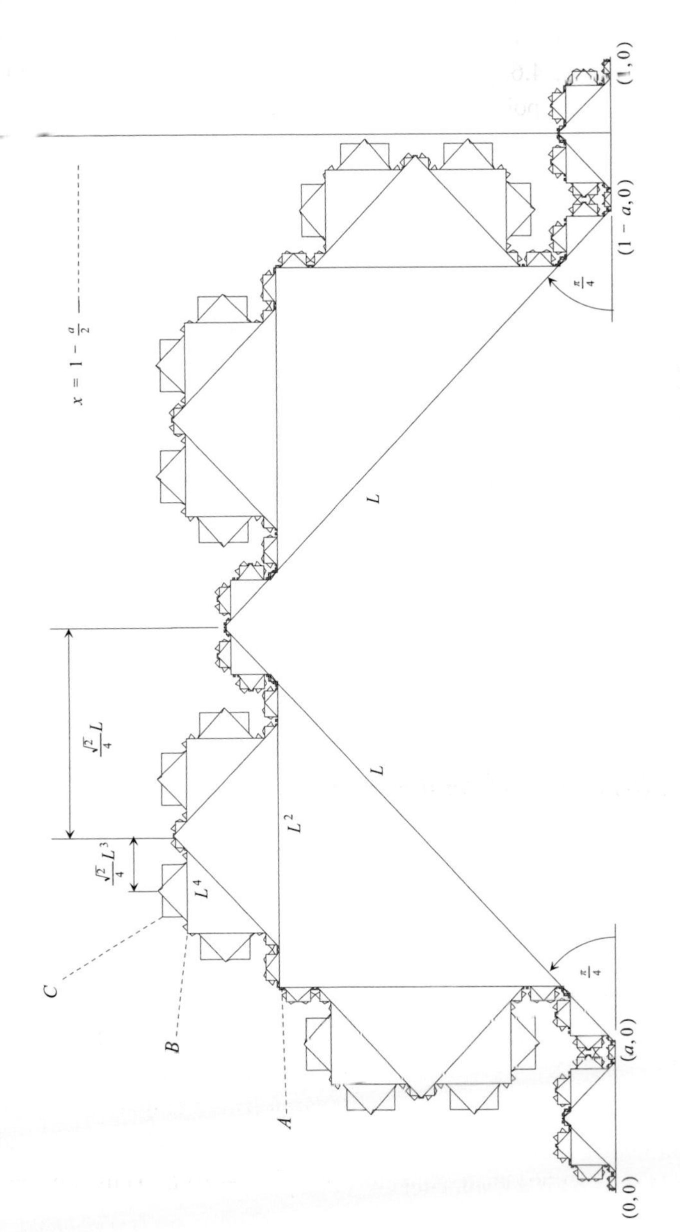

Fig. 4.7: Graphs of $f_k(I)$ for pivotal value $a(\pi/4)$.

$$A = \left(a + \frac{\sqrt{2}}{2}aL, \frac{\sqrt{2}}{2}(1-a)L\right)$$
$$B = (1 + L^2)\left(a + \frac{\sqrt{2}}{2}aL, \frac{\sqrt{2}}{2}(1-a)L\right)$$
$$C = (1 + L^2 + L^4)\left(a + \frac{\sqrt{2}}{2}aL, \frac{\sqrt{2}}{2}(1-a)L\right)$$

Substituting the value for $H_a\left(1-L^2\right)$ for $\frac{\sqrt{2}}{2}(1-a)L$ as suggested by (4.6) into the above equation yields

$$\frac{H_a}{1-a}=\frac{1}{2}. \tag{4.7}$$

Next, dividing the equality (4.6) by $1-a$ and simplifying gives

$$\frac{\sqrt{2}L}{1-L^2}=1 \tag{4.8}$$

which has positive solution $L=\frac{\sqrt{3}-1}{\sqrt{2}}$. Coupling this with the fact that $L=\frac{\sqrt{2}}{2}(1-2a)$ we obtain that the pivotal value is $a=a(\pi/4)=1-\frac{\sqrt{3}}{2}$. Formula (4.7) then yields $H_{1-\sqrt{3}/2}=\frac{1}{2}\frac{\sqrt{3}}{2}=\frac{\sqrt{3}}{4}$ for the height of the curve.

4.3.2 *Double points form Cantor sets*

Recall that the function $f\equiv f_a$ depends on a. To examine the double points of this curve we need to define a class of Cantor sets determined by the function f in equation (4.3). We begin by considering a generic $a\in(0,1/2)$ and focusing on the intersection of $f_a(I)$ with I. Note that $f_a(I)\cap I=f_a(S)$, where

$$S=\{t=0.t_1t_2t_3\cdots\text{base}\,4:t_j\in\{0,3\},\,j\geq 1\}. \tag{4.9}$$

For $t\in S$ we have

$$f_a(t)=(1-a)\sum_{j=1}^{\infty}a^{j-1}\frac{t_j}{3}.$$

The map $f_a:S\to f_a(S)=f_a(I)\cap I$ is a one-to-one, continuous, increasing map.

When $a=1/3$, we have

$$\begin{aligned}f_{1/3}(t)&=\frac{2}{3}\sum_{j\geq 1}\left(\frac{1}{3}\right)^{j-1}\frac{t_j}{3}\\&=\sum_{j\geq 1}\left(\frac{1}{3^j}\right)\frac{2t_j}{3}=(0.2t_1/3,2t_2/3\cdots\text{base}\,3),\end{aligned}$$

so $f_{1/3}(S)$ is the standard Cantor set C. Consequently, the map $f_a\circ(f_{1/3})^{-1}$ is a one-to-one, continuous, increasing map from C onto $f_a(S)$. Hence $f_a(S)$ is homeomorphic to C so $f_a(S)$ is a Cantor set.

Put $\widetilde{C}_a=f_a(S)$. For our purposes in this section, *a Cantor set is a reduced, rotated, and translated copy of some* $\widetilde{C}_a$. A point $t\in\mathbb{R}$ is an *endpoint* of $\widetilde{C}_a$ if and only if $t\in\widetilde{C}_a$ and t is a boundary point of some component of $\mathbb{R}-\widetilde{C}_a$.

We continue with $a = 1 - \sqrt{3}/2$. From the drawing in Figure 4.6 it appears that the top $T = \{(x, y) : y = H_a\}$ is a Cantor set and that a point $t = 0.t_1t_2 \cdots \text{base}\, 4$ in I maps onto T if and only if each pair $\{t_{2j-1}, t_{2j}\}$ is composed of one 1 and one 2. If we calculate the x-coordinate x_l of the left endpoint of T, then we can use symmetry to find the other endpoints of T.

Tracing the regions of the curve as we did earlier with the RS curve (See Section 4.2 of this chapter.), we see that the left endpoint of T is the image of $t = 0.\underline{12}\,\text{base}\,4$. From (4.5) we have $f\,(0.\underline{12}\,\text{base}\,4) = \left(1/4, \sqrt{3}/4\right)$ so $x_l = 1/4$. Since $f\,(1 - x) = 1 - f\,(x)$ for $x \in I$, the right endpoint of T is $x_r = 1 - x_l = 3/4$. Finally, the right endpoint of the left half of T is located at $f\,(0.12\underline{21}\,\text{base}\,4)$ and has x-coordinate $\left(1/4 - \sqrt{3}/2\right) = 1/4 + a$. Hence, T is a $1/2$-sized copy of $\widetilde{C}_{2a} = \widetilde{C}_{2-\sqrt{3}}$.

The intersection of T_{022} and the line $x = a(1 - a/2)$ is a L^2-sized copy of T which has been rotated $\pi/2$ clockwise and suitably translated. This set of double points is composed of images of points $t = 0.022t_3t_4 \cdots t_{2k-1}t_{2k} \cdots \text{base}\,4$, where each pair $t_{2k-1}t_{2k}$ is composed of one 1 and one 2. Thus, this intersection is a Cantor set on the line $x = a(1 - a/2)$; moreover, a point p in I maps onto this intersection if and only if $p = 0.022 + t/64$ where t a number that maps onto T. The endpoints of this Cantor set are boundary points of lagoons surrounded by K_a. These lagoons all have the same shape. The image in Figure 4.8 is an enlargement of a small piece of Figure 4.6 where you can see an approximation to some of these lagoons. We will see in Chapter 5 that the sequence $\{f_k(I)\}$ of images converges to K_a in the Hausdorff metric. As k increases, more lagoons appear and begin to take shape.

You may want to find the pivotal values and explore the behavior of other curves derived with different angles. We suggest the angles $\theta = \pi/5$ and $\theta = \pi/6$ to note the remarkable difference between the characterizations of the double points in these two cases.

4.4 Problems

PROBLEM 4.1. Let $a = 1/4$ and $t \in S = \{t = 0.t_1t_2t_3 \cdots : t_j \in \{0, 3\},\ j \geq 1\}$. Show that $S = \widetilde{C}_{1/4}$.

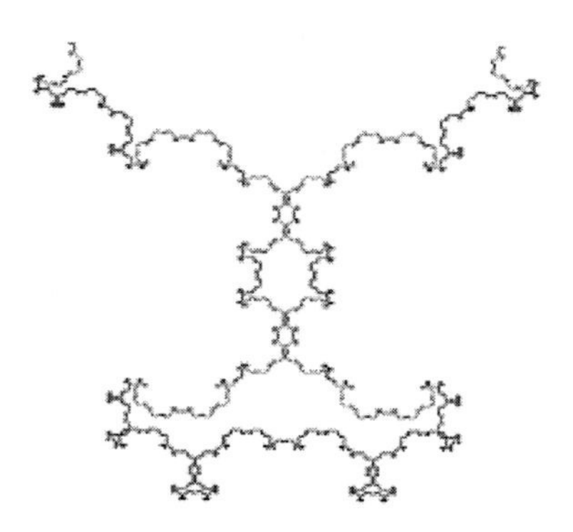

Fig. 4.8: Lagoons of the curve with $\theta = \pi/4$.

Chapter 5

Metric spaces and the Hausdorff metric

This chapter contains mathematical background for the remainder of the book. A metric on a set S is a general concept of distance between members of S. The Hausdorff metric h is used to define a positive distance between two compact sets in the plane. We will verify that h is a complete metric on the set $\mathcal{K}$ of compact subsets of the plane. The fact that the set of connected sets in $\mathcal{K}$ comprises a closed set in the complete metric space $(\mathcal{K}, h)$ encourages the use of Iterated Function Systems to generate curves as we do in Chapter 6. We begin this chapter with two examples of metrics on the plane to expand your thinking beyond the Euclidean metrics.

5.1 Metric spaces

Definition 5.1. A **metric space** is a set S together with a (distance) map d from $S \times S$ to $\mathbb{R}$ with the following properties for any points x, y and z in S:

(1) $d(x, x) = 0$,
(2) $d(x, y) = d(y, x) > 0$, for $x \neq y$, and
(3) $d(x, z) \leq d(x, y) + d(y, z)$ (triangle inequality).

We often designate this metric space by (S, d).

Example 5.1 (The Cell metric in a totally connected wireless world). Let S be a set of people, each of whom has a "universal cell phone": any person in S can call any other person in S. The (calling) distance $d_{\text{cell}}(x, y)$ between any two people x and y in S is equal to 1; the distance $d_{\text{cell}}(x, x) = 0$ for each person x. (Each person is alone in a cell; any two people are one call apart.)

Example 5.2. Let S be a set of accessible places in the world; that is, there is at least one practical path between any two places in S (no breaking through walls, for example). The distance $d_{\text{arc}}(x, y)$ between any two points x and y in S is the infimum of the lengths of all practical paths from x to y. The distance between two places is greater than zero and the distance from a place to itself is equal to zero. Also, the triangle inequality is valid since joining any practical path from x to y with any practical path from y to z is a practical path from x to z.

Definition 5.2. Let $z = (x, y)$ and $w = (u, v)$ be points in $\mathbb{R}^2$. The functions d_m, d_2, and d_1 defined on $\mathbb{R}^2 \times \mathbb{R}^2$ by

$$d_m(z, w) = \max\{|x - u|, |y - v|\},$$

$$d_2(z, w) = \sqrt{(x - u)^2 + (y - v)^2},$$

and

$$d_1(z, w) = |x - u| + |y - v|$$

are metrics on $\mathbb{R}^2$.

Examples 5.1 and 5.2 are modified below to give two metrics on $\mathbb{R}^2$.

Example 5.3 (A Cell Metric on $\mathbb{R}^2$). The metric d_{cell} on $\mathbb{R}^2$ is defined by $d_{\text{cell}}(z, w) = 1$ if $z \neq w$.

Example 5.4 (A Corral Metric on $\mathbb{R}^2$). Referring to Figure 5.1, we consider a closed arc of 330 degrees on the unit circle. Think of the unit circle as a corral in the plane with the part of the circle that is not in the arc as the entrance to the corral. The arc can be touched from the inside, but not from the outside. *Admissible paths* between a point z in $\mathbb{R}^2$ and a point w in $\mathbb{R}^2$ are the polygonal paths between z and w which do not touch the outside of the arc. The metric d_{arc} is defined on $\mathbb{R}^2$ by the formula

$$d_{\text{arc}}(z, w) = \inf\{\text{lengths of admissible paths between } z \text{ and } w\}.$$

Problem 5.1 asks you to verify that d_{cell}, d_{arc}, d_m, d_2, and d_1 are metrics.

5.1.1 *Equivalent metrics*

Two metrics d_a and d_b are said to be equivalent metrics on a set S if and only if there are positive numbers P_a and P_b such that for each x and y in S,

$$d_a(x, y) \leq P_a d_b(x, y) \text{ and } d_b(x, y) \leq P_b d_a(x, y). \tag{5.1}$$

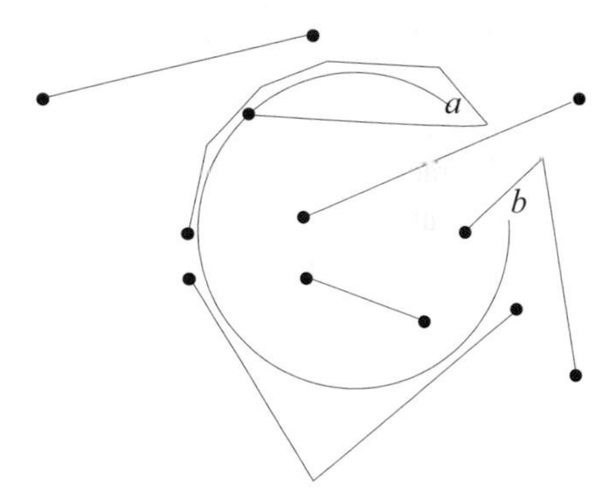

Fig. 5.1: Acceptable polygonal paths between points in the plane. Note that any polygonal path that connects the endpoints a and b and otherwise does not intersect the circular arc is acceptable.

The metrics d_1, d_2, and d_m for $\mathbb{R}^2$ satisfy

$$d_m(z,w) \leq d_2(z,w) \leq d_1(z,w) \leq 2d_m(z,w) \tag{5.2}$$

so they are equivalent.

Further equivalence properties of $d_{\text{cell}}, d_{\text{arc}}, d_m, d_2$, and d_1 are examined in Problem 5.2.

5.1.2 *Topological properties of metric spaces*

Topological properties of objects in a metric space can be formulated in the terminology of its metric. Suppose throughout that (S,d) is a metric space.

Definition 5.3. For $x \in S$ and r a positive number, the **open ball** $B(x,r)$ of radius r about x (in (S,d)) is

$$B(x,r) = \{w \in S : d(x,w) < r\}.$$

For $E \subset S$ and r a positive number, the **open ball** $B(E,r)$ of radius r about E (in (S,d)) is

$$B(E,r) = \bigcup_{x \in E} B(x,r).$$

Definition 5.4. A subset E of a metric space (S,d) is an **open set** (in S) if, for each point $x \in E$, there is a positive number r_x such that $B(x,r_x) \subseteq E$.

An arbitrary union of open sets is an open set, and a finite intersection of open sets is an open set. (See Problem 5.3.)

Proposition 5.1. *In* $\mathbb{R}^k$, $k = 1, 2$, *an arbitrary union of open sets is equal to a countable union of those open sets.*

Proof. We will prove the result for $k = 2$. The case for $k = 1$ is similar. Let

$$V = \bigcup_{\alpha \in \Omega} O_\alpha$$

be an arbitrary union of open sets O_α. For each $x \in V$, choose a point r_x in $\mathbb{R}^2$ with rational coordinates and a positive integer n_x such that $x \in B(r_x, 1/n_x)$ is contained in at least one O_α; choose one such α and label it $\alpha(r_x, n_x)$. There are only countably many pairs (r_x, n_x). For each pair (r_x, n_x), choose one of the indices $\alpha(r_x, n_x)$ corresponding to that pair. Denote the chosen index by $\widetilde{\alpha}(r_x, n_x)$. Only countably many indices $\widetilde{\alpha}(r_x, n_x)$ are chosen. Thus,

$$V = \bigcup_{\widetilde{\alpha}(r_x, n_x)} O_{\widetilde{\alpha}(r_x, n_x)}$$

is a countable union of open sets $O_{\widetilde{\alpha}(r_x, n_x)}$. □

Definition 5.5. A subset K of (S, d) is a **closed set** in (S, d) if whenever $\{w_k\}$ is a sequence of points in K and $\{w_k\}$ converges in (S, d) to a point w, then $w \in K$.

Proposition 5.2. *A subset* M *of a metric space* (S, d) *is an open set (in* S*) if and only if the complement* $S \backslash M$ *of* M *is a closed subset of* S.

Proof. It suffices to verify that M is not closed if and only if $S \backslash M$ is not open. First suppose M is not closed. Then there exists a convergent sequence $\{x_k\}$ of points in M such that $x_k \to x \in S \backslash M$. Consequently, $S \backslash M$ is not open.

Next suppose $S \backslash M$ is not open. Then there exists a point x in $S \backslash M$ such that for each $\epsilon > 0$, the ball $B(x, \epsilon)$ contains points of M. We define a sequence $\{x_k\}$ of points in M that converges to x. Let

$$x_1 \in M \cap B(x, 1);$$

and for $k > 1$, let

$$x_{k+1} \in M \cap B\left(x, \frac{1}{2} d(x, x_k)\right).$$

Hence,

$$d(x, x_k) < \frac{1}{2} d(x, x_{k-1}) \leq \cdots \leq \frac{1}{2^{k-1}} d(x, x_1)$$

so the real sequence $\{d(x, x_k)\}$ converges to 0 suggesting that $\{x_k\}$ converges to x. Thus M is not closed. □

An arbitrary intersection of closed sets is a closed set. And a finite union of closed sets is a closed set. (See Problem 5.4.)

Remark 5.1. In $(\mathbb{R}, d_1)$ the union

$$\cup_{n \in \mathbb{N}}[1/n, 1] = (0, 1]$$

of the closed intervals $[1/n, 1]$ is not closed.

Definition 5.6. A sequence $\{w_k\}$ of elements of S **converges** in (S, d) to an element w of S if and only if for each $\epsilon > 0$, there exists a positive integer K_ϵ such that $k > K_\epsilon$ implies that $w_k \in B(w, r)$ whenever $k > K_\epsilon$.

Note that a sequence $\{w_k\}$ in S converges to w if and only if the real sequence $\{d(w_k, w)\}$ converges to 0.

Definition 5.7. A sequence $\{w_k\}$ is a **Cauchy sequence** in (S, d) if for each $\epsilon > 0$, there exists a positive integer K_ϵ such that $j > K_\epsilon$ and $k > K_\epsilon$ imply that $d(w_j, w_k) < \epsilon$.

Definition 5.8. A point $x \in S$ is a **limit point** of a subset E of S when each ball $B(x, \epsilon)$ contains a point of E distinct from x. In this case there is a sequence of distinct points in E that converges to x.

Definition 5.9. Given $E \subseteq S$ the **closure** $\overline{E}$ of E is the smallest closed set that contains E. Consequently, $\overline{E}$ is the intersection of all closed sets that contain E. In fact, $\overline{E}$ is composed of the points in E together with all limit points of E.

Definition 5.10. A subset K of (S, d) is **bounded** if there exist $w \in S$ and $r > 0$ such that $K \subset B(w, r)$.

Definition 5.11. A subset K of (S, d) is **compact** if whenever $\{w_k\}$ is a sequence of points in K, then some subsequence $\{w_{k_j}\}$ of $\{w_k\}$ converges in (S, d) to a point w in K.

Equivalent metrics preserve the topological properties open, closed, convergence, Cauchy, limit point, closure, bounded, compact, etc.

Definition 5.12. A metric space in which each sequence has a convergent subsequence is called a **compact metric space**.

Definition 5.13. For a fixed $r > 0$ the **closed ball** $B^c(x, r)$ is defined by

$$B^c(x, r) = \{w \in S : d(x, w) \leq r\}.$$

For any E a subset of a metric space (S, d)

$$B^c(E, r) = \cup_{x \in E} B^c(x, r).$$

is the **closed ball about** E of radius r.

Notice that $B^c(x,r)$ is a closed subset of S. Then, in $(\mathbb{R}^2, d_2)$, for $x \in S$ and $r > 0$, $B^c(x,r) = \overline{B(x,r)}$. However, for $x \in (\mathbb{R}^2, d_{\text{cell}})$

$$\overline{B(x,1)} = \{x\} \text{ and } B^c(x,1) = \mathbb{R}^2.$$

In Problem 5.6 you are asked to verify that if $\mathbb{R}^2$ is equipped with any of the metrics d_m, d_1 and d_2, then a set is compact if and only if it is closed and bounded. The space $(\mathbb{R}^2, d_{\text{cell}})$ is an example of a metric space in which closed and bounded is not equivalent to compact. Recall that (Problem 5.2) the metrics d_{cell} and d_{arc} are examples of metrics on $\mathbb{R}^2$ that are not equivalent to the Euclidean metric.

Definition 5.14. Let E be a subset of a metric space (S,d) and let (T,d^*) be another metric space. The map $m : E \to (T,d^*)$ is **continuous** at $x \in E$ if for every positive number ϵ there is a positive number δ such that $d(w,x) < \delta$, with $w \in E$, ensures that $d^*(m(w), m(x)) < \epsilon$. A map is continuous on E if it is continuous at every point of E.

Example 5.5. Let d_1 and d_{cell} be the previously defined metrics on $\mathbb{R}^2$ and let $f(x) = x$ for $x \in \mathbb{R}^2$. Then the function f as a map from $(\mathbb{R}^2, d_1)$ to $(\mathbb{R}^2, d_{\text{cell}})$ is nowhere continuous. To verify this choose $\epsilon = 1/2$ so that regardless of how small $d_1(x,y) \neq 0$ is, $d_{\text{cell}}(x,y) > \epsilon$.

On the other hand if f is thought of as a map from $(\mathbb{R}^2, d_{\text{cell}})$ to $(\mathbb{R}^2, d_1)$, then f is continuous. (Recall that the two metrics used in this example are not equivalent. The function f is continuous on $\mathbb{R}^2$ if two equivalent metrics are used. See Problem 5.11.)

Definition 5.15. Let $E \subset \mathbb{R}^2$. The collection of open sets $\{O_\alpha\}_{\alpha\in\Omega}$ **covers** E if $E \subset \cup_{\alpha\in\Omega} O_\alpha$ in which case the collection $\{O_\alpha\}_{\alpha\in\Omega}$ is called a **open cover** of E. According to Proposition 5.1 we may assume that the indexing set Ω contains at most a countable number of elements.

Definition 5.16. A **finite subcover** of an open cover $\{O_\alpha\}_{\alpha\in\Omega}$ of a subset $E \subset \mathbb{R}^2$ is a subset $\{O_{\alpha_1}, \ldots, O_{\alpha_k}\}$ of $\{O_\alpha\}_{\alpha\in\Omega}$ for which $E \subset \cup_{j=1}^k O_{\alpha_j}$.

Proposition 5.3. *If E is a compact set in $\mathbb{R}^2$ and $U = \{O_j\}_{j\in\mathbb{N}}$ is an open cover of E then U contains a finite subcover of E.*

Proof. For $k \geq 1$, put

$$E_k = E \setminus \bigcup_{j\leq k} O_j.$$

Since each E_k is closed and bounded it is compact. Also, $E_1 \supseteq E_2 \supseteq \cdots$ and $\bigcap_{k\geq 1} E_k$ is empty because $\cup_{j\in\mathbb{N}} O_j$ covers E. According to Proposition C.7 there is a least positive integer k for which E_k is empty. Thus, $\bigcup_{j\leq k} O_j$ is a finite *subcover* of E. □

5.1.3 *Complete metric spaces*

Definition 5.17. A metric space (S, d) is **complete** if each Cauchy sequence in (S, d) converges to an element of S.

The metrics d_{cell} and d_{arc} were introduced as examples of metrics on $\mathbb{R}^2$ that are not equivalent to the Euclidean metric. The metric space $(\mathbb{R}^2, d_{\text{arc}})$ is not complete, and $(\mathbb{R}^2, d_{\text{cell}})$ is an example of a complete metric space where closed and bounded is not equivalent to compact.

Definition 5.18. Let E be a subset of a metric space (S, d). For all $x \in S$ the **distance** between x and E is the number

$$d(x, E) = \inf \{d(x, y) : y \in E\}.$$

The next example illustrates distance on metric spaces.

Example 5.6. Put $E = (2, 3]$, $F = [0, 1]$, $S = F \cup E$, $d(x, y) = 1$ when $2 < x < y \leq 3$ and $d(x, y) = |x - y|$ when $x \in F$. (We have the cell metric on E and the Euclidean distance when at least one of the points is in F.) Then (S, d) is a complete metric space and E is a non-compact, closed and bounded subset of S with

$$\bigcup_{x \in E} B^c(x, 1) = (0, 1] \cup (2, 3],$$

which is not closed because

$$\overline{\bigcup_{x \in E} B^c(x, 1)} = S.$$

Moreover, $d(1, E) = 1$, but $d(1, y) > 1$ if $y \in E$.

The following propositions show that such behavior could not occur if E were compact.

Proposition 5.4. *Let (S, d) be a complete metric space. If K is a compact subset of S and $x \in S$ then there exists $y \in K$ for which*

$$d(y, x) = d(K, x).$$

Proof. Apply the proof of Proposition C.8. □

Proposition 5.5. *Let (S, d) be a complete metric space. If K is a compact subset of S and $\epsilon > 0$, then $B^c(K, \epsilon)$ is a closed set.*

Proof. Let p be a point in the closure $\overline{B^c(K,\epsilon)}$ of $B^c(K,\epsilon)$. For $n \in \mathbb{N}$ choose $z_n \in B^c(K,\epsilon)$ with $d(p,z_n) < 1/n$. According to Proposition 5.4 for $n \in \mathbb{N}$ we can choose $s_n \in S$, with $d(z_n,s_n) \leq \epsilon$. Then some subsequence $\{s_{n_k}\}$ of $\{s_n\}$ converges to a point $s \in S$. Hence,

$$\begin{aligned} d(p,s) &\leq d(p,z_{n_k}) + d(z_{n_k},s_{n_k}) + d(s_{n_k},s) \\ &\leq d(p,z_{n_k}) + \epsilon + d(s_{n_k},s) \to \epsilon \end{aligned}$$

so

$$p \in B^c(K,\epsilon).$$

□

5.2 The Hausdorff metric

The Hausdorff metric, named for Felix Hausdorff, is our next example of a metric. This metric measures the distance between nonempty compact subsets of a metric space. In recent years the Hausdorff metric has played an important role in the development of fractal geometry. We will use it in Chapter 6 for our discussion of fractals. Let (X,d) be a complete metric space and let $\mathcal{K} = \mathcal{K}(X)$ denote the space whose points are the nonempty compact subsets of X.

Example 5.7. If $X = \mathbb{R}^2$, then $\mathcal{K}$ is all nonempty closed and bounded subsets of $\mathbb{R}^2$.

We define the **Hausdorff metric** h on the set $\mathcal{K} = \mathcal{K}(\mathbb{R}^2)$ of nonempty compact subsets of $\mathbb{R}^2$ in terms of the Euclidean metric $d = d_2$ on $\mathbb{R}^2$. The Hausdorff metric compares the shapes of elements of $\mathcal{K}$. Referring to Appendix C, for $z \in \mathbb{R}^2$ and $E, F \in \mathcal{K}$ we have

$$d(z,F) = \min_{w \in F} d(z,w)$$

and

$$d(E,F) = \min\{d(x,y) : x \in E, y \in F\}.$$

The function d measures the Euclidean distance between two compact sets; it is NOT a metric since if $E \cap F \neq \emptyset$, then $d(E,F) = 0$. The Hausdorff distance is a metric on $\mathcal{K}$.

Definition 5.19. The **Hausdorff distance** $h(E,F)$ between E and F in $\mathcal{K}$ is

$$h(E,F) = h(F,E) = \inf\{p : E \subset B(F,p) \text{ and } F \subset B(E,p)\}.$$

Example 5.8. Suppose $X = \mathbb{R}$, $x = 5$, $A = [-4, -1]$ and $B = [1, 7]$. Then $d(x, A) = 6$, $d(A, B) = 5$, $d(B, A) = 8$ and $h(A, B) = 8$.

A list of properties regarding d, h, and $\mathcal{K}$ follows. The proofs are left as exercises for you. We will use these properties to establish Proposition 5.6 about nested sequences in $\mathcal{K}$ and to show that $(\mathcal{K}, h)$ is a complete metric space. In the following properties we assume that E, F and G are subsets of $\mathbb{R}^2$ and a and b are positive numbers.

Property 1. If $E \subset B(F, a)$, then $B(E, b) \subset B(F, a + b)$.

Property 2. If $E \subset B(F, a)$ and $F \subset B(G, b)$, then $E \subset B(G, a + b)$.

Property 3. h is a metric on $\mathcal{K} = \mathcal{K}(\mathbb{R}^2)$.

Property 4. For $E \in \mathcal{K}$, $F \in \mathcal{K}$, and $F \subset E$, $h(E, F) = \max_{z \in E} d(z, F)$.

The following is an extension of Proposition C.7.

Proposition 5.6. *Let $\{F_k\}$ be a nonincreasing sequence of elements of $\mathcal{K}$; that is, $F_{k+1} \subset F_k$. Then $\bigcap_{k \geq 1} F_k = F \in \mathcal{K}$. Moreover, $h(F, F_k)$ decreases monotonically to zero:*

$$h(F, F_k) \searrow 0.$$

Proof. According to Proposition C.7

$$\bigcap_{k \geq 1} F_k = F \in \mathcal{K}.$$

To show that the sequence $h(F, F_k)$ decreases monotonically to zero, notice first that Property 4 implies that the sequence $\{h(F, F_k)\}$ is non-increasing. Next, applying Property 4 again, for each positive integer k, there exists $z_k \in F_k$ such that $d(z_k, F) = h(F_k, F)$. A subsequence $\{z_{k_j}\}$ of the sequence $\{z_k\}$ converges to a point $z \in F$. Thus,

$$h(F_{k_j}, F) = d(z_{k_j}, F) \leq d(z_{k_j}, z) \longrightarrow 0.$$

□

Remark 5.2. Proposition 5.6 says that the shape of F_k is a uniformly close approximation to the shape of F for k sufficiently large: given $\epsilon > 0$, there is a positive integer n_ϵ such that if $d(z, F) > \epsilon$, then $z \notin F_{n_\epsilon}$. That is, $F \subset F_{n_\epsilon} \subset B^c(F, \epsilon)$.

Theorem 5.1. *The space $(\mathcal{K}, h)$ is a complete metric space.*

Proof. Suppose that $\{E_k\}$ is a Cauchy sequence in $\mathcal{K}$. Choose an increasing subsequence $\{n_k\}$ of positive integers such that

$$h(E_i, E_j) < 2^{-(k+1)}$$

if $i, j \geq n_k$. It suffices to show that the subsequence $\{E_{n_k}\}$ converges. We restrict our attention to this subsequence, and relabel n_k as k, to simplify the notation.

After the preceding modification, we establish the theorem by showing that the sequence $\{E_k\}$ converges when

$$h(E_k, E_m) < 2^{-(k+1)}$$

for $m > k$. Below, we modify the sequence $\{E_k\}$ slightly to obtain a non-increasing sequence $\{F_k\}$. For $k \geq 1$, put

$$F_k = B^c(E_k, 2^{-k}).$$

By Proposition 5.5 F_k is a closed subset of $\mathbb{R}^2$. The set F_k is compact because it is also bounded. We have

$$h(E_k, F_k) = 2^{-k}.$$

Moreover,

$$\begin{aligned} F_{k+1} = B^c(E_{k+1}, 2^{-(k+1)}) &\subset B^c\left(B^c(E_k, 2^{-(k+1)}), 2^{-(k+1)}\right) \\ &= B^c(E_k, 2^{-k}) = F_k. \end{aligned}$$

Let $F = \cap F_k$. According to Proposition 5.6,

$$h(F, F_k) \searrow 0.$$

Because $h(E_k, F_k) \to 0$,

$$h(E_k, F) \to 0.$$

□

The metric space $(\mathcal{K}, h)$ is discussed in Chapter 6.

5.3 Metrics and norms

A norm is a function that is related to a special type of metric. We describe this relationship for the setting $\mathbb{R}^2$, corresponding to a metric d on $\mathbb{R}^2$ with the **translation property**

$$d(w + z, z) = d(w, 0)$$

and the **triangle property**

$$d(w + z, 0) \leq d(w, 0) + d(z, 0).$$

We define a function called a **norm** $\|\cdot\|$ by the equation

$$\|z\| = d(z, 0).$$

This norm on $\mathbb{R}^2$ has three properties:

Positivity: $\|z\| > 0 \Leftrightarrow z \neq 0$,

The triangle inequality: $\|w + z\| \leq \|w\| + \|z\|$, and

Homogeneity: $\|az\| = |a| \, \|z\|$

for all $a \in \mathbb{R}$ and $z, w \in \mathbb{R}^2$.

Conversely, beginning with a norm $\|\cdot\|$, we can define a corresponding metric d with the translation property and the triangle property by the equation

$$d(w, z) = \|w - z\|.$$

Example 5.9. Define a metric $d_{\text{sqrt}}(u, v)$ for $u = (u_1, u_2)$ and $v = (v_1, v_2)$, by the equation

$$d_{\text{sqrt}}(u, v) = \sqrt{|u_1 - v_1|} + \sqrt{|u_2 - v_2|},$$

and recall the five metrics d_1, d_2, d_m, d_{cell}, and d_{arc}. Among these six metrics, all but d_{arc} have the triangle property, but only d_1, d_2, and d_m, have the homogeneity property. You should verify that d_{sqrt} is equivalent to none of the other five metrics.

Thus, we see that metrics can be very dissimilar. However, such dissimilarity does not occur among norms; any two norms on $\mathbb{R}^2$ are equivalent and the corresponding metrics are equivalent.

Theorem 5.2. *Any two norms on $\mathbb{R}^2$ are equivalent.*

Proof. Perhaps the easiest way to verify this important fact is to show that any norm on $\mathbb{R}^2$ is equivalent to the norm $\|u\|_1 = |u_1| + |u_2|$. Let $\|\cdot\|$ denote any norm on $\mathbb{R}^2$. For $z = (x, y) \in \mathbb{R}^2$ we have

$$\begin{aligned} \|z\| = \|(x, y)\| &\leq |x| \, \|(1, 0)\| + |y| \, \|(0, 1)\| \\ &\leq \max\{\|(1, 0)\|, \|(0, 1)\|\} \, \|z\|_1 = m \, \|z\|_1 . \end{aligned}$$

To verify that a positive number k exists with the property that for all $z \in \mathbb{R}^2$ we have $\|z\|_1 \leq k \, \|z\|$, we suppose, on the contrary, that there is a sequence $\{z_k\}$, where, using homogeneity, we have

$$\|z_k\|_1 > k \, \|z_k\| \text{ or } 1/k > \frac{\|z_k\|}{\|z_k\|_1} = \left\| \frac{z_k}{\|z_k\|_1} \right\| .$$

We arrive at a contradiction by setting

$$w_k = (u_k, v_k) = \frac{z_k}{\|z_k\|_1}.$$

We now have $\|w_k\|_1 = 1$, so some subsequence $w_{k_j} \to w$, where $\|w\|_1 = 1$ because $\|w_{kj}\|_1 = 1$. But, we also have $\|w\|_1 = 0$ because $\|w_{kj}\|_1 \leq m \|w_k\| \to 0$. □

In Section C.6 we use $A = A(\mathbb{R}^2)$ to denote the set of linear maps (operators) $M : \mathbb{R}^2 \to \mathbb{R}^2$. Each map M in A has a representation as a 2×2 matrix $M = \begin{bmatrix} a & c \\ b & d \end{bmatrix}$.

Definition 5.20. Define a **norm** on A by

$$\|M\| = \sup\{|Mz| : |z| \leq 1\}.$$

Several properties of this norm appear in the problems. We conclude this section with two pertinent examples.

Example 5.10. To generate the Sierpinski triangle in Example 1.2, we put

$$M = \begin{bmatrix} 1/2 & 0 \\ 0 & 1/2 \end{bmatrix}$$

and let T denote the equilateral triangle with base $I = [0, 1]$. We see that by putting $N = M + (1/2, 0)$ and $O = M + (1/4, \sqrt{3}/4)$ we get three maps that generate the Sierpinski triangle. The set $M(T) \cup N(T) \cup O(T)$ corresponds to the First step in Figure 1.10. Here, $\|M\| = 1/2$.

Example 5.11. The maps

$$M = \frac{1}{\sqrt{2}} M_{-\pi/4} = \begin{bmatrix} 1/2 & 1/2 \\ -1/2 & 1/2 \end{bmatrix}$$

and $N = M + (0, 1)$ generate the *fractile* (tile with fractal boundary [Darst *et al.* (1998)]) discussed in Example 6.5. Here $\|M\| = 1/\sqrt{2}$.

5.4 Problems

PROBLEM 5.1. Show that $d_{\text{cell}}, d_{\text{arc}}, d_m, d_2$, and d_1 are metrics on $\mathbb{R}^2$.

PROBLEM 5.2. Consider the set $\{d_m, d_2, d_1, d_{\text{cell}}, d_{\text{arc}}\}$ of five metrics defined on $\mathbb{R}^2$. Determine which pairs of these metrics are equivalent.

PROBLEM 5.3. Prove that an arbitrary union of open sets is an open set and that a finite intersection of open sets is an open set.

PROBLEM 5.4. Prove that an arbitrary intersection of closed sets is a closed set and that a finite union of closed sets is a closed set.

PROBLEM 5.5. Complete the following problems related to the d_{cell} metric.

(a) Characterize the closed sets in $(\mathbb{R}^2, d_{\text{cell}})$.
(b) Give an example of a closed and bounded set in $(\mathbb{R}^2, d_{\text{cell}})$ which is not compact.
(c) Characterize the compact sets in $(\mathbb{R}^2, d_{\text{cell}})$.
(d) Characterize the Cauchy sequences in $(\mathbb{R}^2, d_{\text{cell}})$.
(e) Show that $(\mathbb{R}^2, d_{\text{cell}})$ is a complete metric space.
(f) For $z \in \mathbb{R}^2$, describe $B(0,1)$ and $B^c(0,1)$ with respect to $(\mathbb{R}^2, d_{\text{cell}})$.

PROBLEM 5.6. Let $S = \mathbb{R}^2$. Suppose that $d = d_m$, d_2, or d_1. Show that a set $K \subset S$ is compact in (S,d) if and only if, K is closed and bounded in (S,d).

PROBLEM 5.7. Let $S = \mathbb{R}^2$. Sketch $B(0,1)$ and $B^c(0,1)$ for each of d_m, d_2, and d_1.

PROBLEM 5.8. Show that if a sequence $\{w_k\}$ converges in (S,d), then $\{w_k\}$ is a Cauchy sequence in (S,d).

PROBLEM 5.9. Suppose that a Cauchy sequence $\{w_k\}$ in (S,d) has a subsequence that converges to an element w of S. Show that $\{w_k\}$ converges to w.

PROBLEM 5.10. Show that if (S,d_a) and (S,d_b) are equivalent metric spaces and $\{w_k\}$ is a convergent sequence in (S,d_a), then $\{w_k\}$ is a convergent sequence in (S,d_b).

PROBLEM 5.11. Let d_1 and d_2 be equivalent metrics on an arbitrary space X. Prove that the identity map $f : (X,d_1) \longrightarrow (X,d_2)$ defined by $f(x) = x$ is continuous for all $x \in X$.

PROBLEM 5.12. Show that $(\mathbb{R}^2, d_{\text{arc}})$ is not a complete metric space.

PROBLEM 5.13. Let $d(x,y) := |x^3 - y^3|$ for $x, y \in \mathbb{R}$. It is easy to show that $(\mathbb{R}, d)$ is a metric space since $d(x,y) = d_1(x^3, y^3)$. Show that d and d_1 are not equivalent metrics on $\mathbb{R}$. Prove that $f : (\mathbb{R}, d_1) \longrightarrow (\mathbb{R}, d)$ defined by $f(x) = x^2$ is continuous.

PROBLEM 5.14. Let (S, d) be a metric space. Let $x \in S$, and let $p > 0$. Show that $B^c(x, p)$ is a closed set.

PROBLEM 5.15. Suppose that A and B are two disjoint compact sets in $\mathbb{R}^2$. Show that there exist points $a \in A$ and $b \in B$ such that $|a - b| = d(A, B)$.

PROBLEM 5.16. Let S be a disconnected, compact subset of $\mathbb{R}^2$. Let $S = T \cup U$, where T and U are disjoint, closed sets. Let $a \in T$ and $b \in U$. Show that if $\epsilon < d(T, U)$, then there does not exist a finite sequence $\{t_j\}_{0 \leq j \leq n}$ of points in S such that $t_0 = a$, $t_n = b$, and $|t_j - t_{j-1}| \leq \epsilon$ for $1 \leq j \leq n$.

PROBLEM 5.17. Show that if T and U are (nonempty) disjoint closed subsets of $\mathbb{R}^2$, and T is compact, then $d(T, U) > 0$.

PROBLEM 5.18. Show that the product $A \times B$ of a compact set A and a compact set B is a compact set.

PROBLEM 5.19. Let S be a compact, connected subset of $\mathbb{R}^2$. Let m be a continuous map from S to $\mathbb{R}^2$. Show that $m(S)$ is a compact, connected set in $\mathbb{R}^2$.

PROBLEM 5.20. Show that a continuous map from a compact subset S of $\mathbb{R}^2$ to $\mathbb{R}^2$ is uniformly continuous on S.

PROBLEM 5.21. Let (M, d) be a metric space. Show that a finite union of compact sets is compact.

PROBLEM 5.22. Let $G \subset \mathcal{K}$. Suppose that G is a bounded subset of $\mathcal{K}$; that is, suppose there exists $F \in \mathcal{K}$ and p, $0 < p \in \mathbb{R}$, such that

$$G \subset B_h(F, p) = \{E \in \mathcal{K} : h(E, F) < p\}.$$

Show that if G is a closed subset of $\mathcal{K}$, then G is a compact subset of $\mathcal{K}$. (Hint: With some care, you can choose p so that $G \subset \mathcal{K}(B_{d_2}(0, p))$.)

PROBLEM 5.23. Verify that h, the Hausdorff distance, is a metric on $\mathcal{K}$.

PROBLEM 5.24. Let E be a nonempty compact subset of $\mathbb{R}^2$. Let $\mathcal{K}(E)$ denote the set of nonempty, closed subsets of E.

(a) Show that $\mathcal{K}(E)$ is a closed subset of $(\mathcal{K}, h)$.
(b) Show that $(\mathcal{K}(E), h)$ is a complete metric space.
(c) Let $e > 0$. Show that there exists a finite subset $S = S_{E,e}$ of $\mathbb{R}^2$ (S generally depends on E and e) with the following property: If F is a subset of E, then there is a subset S_F of S such that $h(F, S_F) < e$.
(d) Show that you can choose a subset S of E which satisfies the property required in part c.
(e) Show that if $\{E_k\}$ is a sequence of elements of $\mathcal{K}(E)$, then $\{E_k\}$ has a convergent subsequence in $(\mathcal{K}(E), h)$. (*Hint: Notice that the set* S *in part d has only finitely many subsets.) The set* $\mathcal{K}(E)$ has a property called **totally bounded** with respect to h.

PROBLEM 5.25. Put

$$B_1 = \{z \in \mathbb{R}^2 : d_1(z, 0) < 1\},$$

$$B_2 = \{z \in \mathbb{R}^2 : d_2(z, 0) < 1\},$$

and

$$B_m = \{z \in \mathbb{R}^2 : d_m(z, 0) < 1\}.$$

For r and $s \in \{1, 2, m\}$, describe the sets

$$H_{r,s} = \{E \in \mathcal{K} : E \subset B_r(B_s, 1/2) = \cup_{z \in B_s} B_r(z, 1/2)\}.$$

(Recall that $B_r(z, 1/2) = \{w \in \mathbb{R}^2 : d_r(w, z) < 1/2\}$.)

PROBLEM 5.26. For $F \in \mathcal{K}$ and $z \in \mathbb{R}^2$, in the metric space $\left(\mathbb{R}^2, d\right)$ show that $d\,(F, z) = \min_{w \in F} d\,(w, z)$.

PROBLEM 5.27. For $E \in \mathcal{K}$, $F \in \mathcal{K}$ and $F \subset E$, show that $h\,(F, E) = \max_{z \in E} d\,(F, z)$.

PROBLEM 5.28. Show that for $M \in A = A(\mathbb{R}^2)$

$$\begin{aligned} \|M\| &= \sup\{|Mz| : |z| = 1\} \\ &= \sup\left\{\frac{|Mz|}{|z|} : z \neq 0\right\}. \end{aligned}$$

PROBLEM 5.29. Show that a map $M \in A(\mathbb{R}^2)$is a uniformly continuous map.

PROBLEM 5.30. For $M, N \in A$, $\|MN\| \leq \|M\|\,\|N\|$.

PROBLEM 5.31. Let $d_A(M, N) = \|M - N\|$. Show that (A, d_A) is a complete metric space.

PROBLEM 5.32. Give an example where $M, N \in A(\mathbb{R}^2)$ and $MN \neq NM$.

PROBLEM 5.33. Suppose some power $M^{(k)}$ of M satisfies $\left\|M^{(k)}\right\| < 1$. Then $\lim_{k\to\infty} \left\|M^{(k)}\right\| = 0$.

PROBLEM 5.34. A vector $z \in \mathbb{R}^2$ is an **eigenvector** for $M \in A$ with corresponding **eigenvalue** $\lambda \in \mathbb{R}$ if

$$Mz = \lambda z.$$

Set $U = \begin{bmatrix} 1.2 & 1.1 \\ 0 & 1.1 \end{bmatrix}$.

(a) Show that U has eigenvalues $\lambda = 1.1$ and $\mu = 1.2$ with corresponding eigenvectors $v = (1.1, -.1)$ and $w = (1, 0)$ by verifying that $Uv = \lambda v$ and $Uw = \mu w$. Note that v and w are linearly independent; that is, they comprise a basis for $\mathbb{R}^2$.

(b) Verify that $U(1, -1) = (.1, -1.1)$ and that $|(.1, -1.1)| < \sqrt{2} = |(1, -1)|$ so that, with $z = (1, -1)$, $|Uz| < |z|$.

Define a new norm $\|\cdot\|_\phi$ on $\mathbb{R}^2$ as follows: For $z = pv + qw \in \mathbb{R}^2$,

$$\|z\|_\phi = \|pv + qw\| = |p| + |q|$$

with p and q arbitrary in $\mathbb{R}^2$.

(c) Verify that $\|\cdot\|_\phi$ is a norm on $\mathbb{R}^2$.

(d) For M in A, put

$$\|M\|_\Phi = \sup\{\|Mz\|_\phi : \|z\|_\phi \leq 1\}.$$

Verify that $\|\cdot\|_\Phi$ is a norm on A.

(e) Now set $M = U^{-1}$. Verify that $Mv = \lambda^{-1}v$ and $Mw = \mu^{-1}w$, with v, w, λ and μ as above.

(f) Show that $\|M\|_\Phi = 1/1.1 < 1$. Compare $|(.1, -1.1)|$ and $|M(.1, -1.1)| = |(1, -1)|$.

(g) Conclude that $\left\|M^{(n)}\right\| \to 0$. Note that $M(.1, -1.1) = (1, -1)$, so you have $|.1, -1.1)| = 1.22 < \sqrt{2} = |M(.1, -1.1)|$.

PROBLEM 5.35. We can write a matrix M in an equivalent form

$$M = \begin{bmatrix} a & c \\ b & d \end{bmatrix} = (a, b, c, d) \in \mathbb{R}^4.$$

(a) Define appropriate versions of d_1, d_2, d_m, and d_{sqrt} for $\mathbb{R}^4$. Then determine the equivalence relations among these metrics.
(b) Define $\|M\|_1 = |a| + |b| + |c| + |d|$. Define appropriate versions of $\|M\|_2$ and $\|M\|_m$.
(c) Let $\langle\cdot\rangle$ denote an arbitrary norm on $\mathbb{R}^4$. Show that $\langle\cdot\rangle$ and $\|.\|_1$ are equivalent norms on $\mathbb{R}^4$.

PROBLEM 5.36. Complete Problem C.27.

Chapter 6

Contraction maps and iterated function systems

In this chapter we consider maps $f : S \to S$, where S is a complete metric space. A fixed point of f is a point $x \in S$ for which $f(x) = x$. The Contraction Mapping Principle states that a contraction map f has a unique fixed point and describes a way to construct a sequence of points in S that converges to the fixed point. We apply this principle to contraction maps defined on the set $\mathcal{K}$ of compact subsets of S, equipped with the Hausdorff metric. We introduce a class of contraction maps from $\mathcal{K}$ to $\mathcal{K}$ that are called iterated function systems (IFS) . We prove that if the fixed point of an IFS is connected, then it is a curve, and we give a condition that is necessary and sufficient for the fixed point to be connected. Examples demonstrating the use of an IFS to generate fractal sets follow.

6.1 Contraction maps

Definition 6.1. Let (S, d) be a metric space. A map $f : S \to S$ is called a **contraction map on** S if there exists a real number $e < 1$ such that $d(f(s), f(t)) \leq ed(s, t)$ for each pair (s, t) in $S \times S$. The smallest e that works is called the **contraction factor** of f.

In Problem 6.1 you are asked to prove that a contraction map is uniformly continuous.

Definition 6.2. Let (S, d) be a metric space. A point x of S is called a **fixed point** of a map $f : S \to S$ if $f(x) = x$.

Example 6.1. Let S be the interval $[1, 2]$ equipped with the Euclidean metric d. Define $f(x) = \sqrt{x} + 0.5$ and observe that $f(x) \in S$ for all $x \in S$. The Mean Value Theorem from elementary calculus implies that f is a contraction map with contraction factor $1/2$ because $f'(1) = 1/2$ and $f'(x) < 1/2$ when $x > 1$.

Elementary algebra shows that the fixed point is $1 + \sqrt{3}/2$. See Figure 6.1.

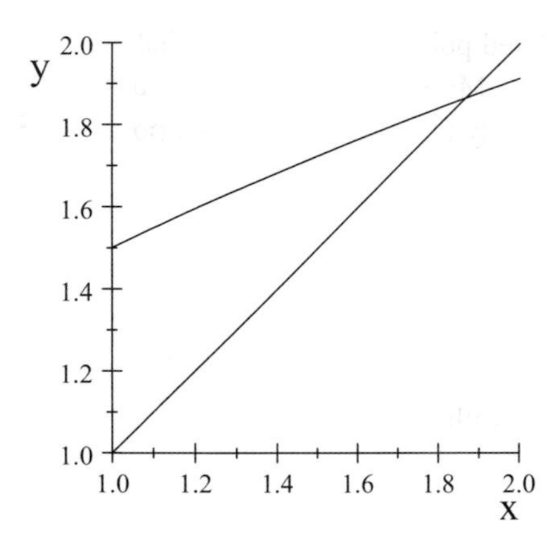

Fig. 6.1: Graphs demonstrating where $f(x) = x$.

Theorem 6.1. Contraction Mapping Principle (Banach Fixed Point Theorem). *Let (S, d) be a complete metric space and let f be a contraction map on S. Then f has a unique fixed point q. Suppose further that x is an element of S and the sequence $\{x_n\}$ is defined by*

$$x_1 = x \text{ and } x_{n+1} = f(x_n), n \geq 1.$$

Then the sequence $\{x_n\}$ converges to q.

Proof. The proof is composed of three steps, A through C below.

A (Find a fixed point q.) Choose an element x in S. Form the sequence $\{x_n\}$ described in the statement of the theorem. Then

$$\begin{aligned} d(x_n, x_{n+1}) &= d\left(f\left(x_{n-1}\right), f\left(x_n\right)\right) \\ &\leq ed(x_{n-1}, x_n) \leq e^2 d(x_{n-2}, x_{n-1}) \\ &\leq \cdots \leq e^{n-1} d(x_1, x_2). \end{aligned}$$

Moreover, for $m = n + k > n$,

$$\begin{aligned} d(x_n, x_m) &\leq d(x_n, x_{n+1}) + d(x_{n+1}, x_{n+2}) + \cdots + d(x_{m-1}, x_m) \\ &\leq (1 + e + e^2 + \cdots + e^{k-1}) d(x_n, x_{n+1}) \\ &\leq \frac{e^{n-1}}{1 - e} d(x_1, x_2). \end{aligned}$$

Thus, the sequence $\{x_n\}$ is a Cauchy sequence in S and it converges to a point in S because (S, d) is a complete metric space. Denote $q = \lim x_n$; $q \in S$.

B (Show that q is a fixed point of f.) Observe that $x_n \to q$ implies that $x_{n+1} = f(x_n) \to f(q)$. Moreover, because x_{n+1} also converges to q, $f(q) = q$.

C (Show that q is the only fixed point of f.) Suppose $u \in S$ and $u \neq q$. Then

$$d(f(u), q) = d(f(u), f(q)) \leq ed(u, q) < d(u, q);$$

thus, $f(u) \neq u$.

□

Remark 6.1. Let f be a contraction mapping on S with fixed point q and contraction factor $e < 1$. Start with any $u \in S$ and let $u_n = \overbrace{f \circ \cdots \circ f}^{n}(u)$. Then

$$d(u_n, q) \leq e^n d(u, q)$$

so the sequence $\{u_n\}$ converges to q since $e^n d(u, q)$ converges to 0. Observe that f pulls u toward the fixed point q at a rate at least equal to e. This means that the sequence $\{u_n\}$ converges **geometrically** to q.

Definition 6.3. Let D be a closed set in any metric space (S, d). A **similarity** $T : D \to D$ is a contraction with

$$d(T(x), T(y)) = ed(x, y)$$

for all x, y in D. In $\mathbb{R}$ and $\mathbb{R}^2$ a similarity T transforms sets into geometrically similar ones.

Example 6.2. The function f given in Example 6.1 is not a similarity since, for example, $d(f(1.5), f(1)) = \sqrt{1.5} - 1 < \frac{1}{4} = \frac{1}{2}d(1.5, 1)$. However, the function $g(x) = mx + b$, $|m| < 1$, is a similarity on $\mathbb{R}$ since for any real numbers x and y we have

$$d(g(x), g(y)) = |mx - my| = |m| \, d(x, y).$$

Recall that $\mathcal{K} = \mathcal{K}(\mathbb{R}^2)$ denotes the set of non-empty, compact subsets of $\mathbb{R}^2$, and h denotes the Hausdorff metric on $\mathcal{K}$. Also recall that the continuous image of a compact set is a compact set and that a continuous map defined on a compact set is uniformly continuous.

Definition 6.4. Let f be a continuous map from $\mathbb{R}^2$ to $\mathbb{R}^2$. The **extension** of f to a map $f : \mathcal{K} \to \mathcal{K}$ is given by

$$f(E) = \{f(x) : x \in E\} \in \mathcal{K} \text{ for } E \in \mathcal{K},$$

where for simplicity we use the same notation for both functions.

Proposition 6.1. *Let d denote the Euclidean metric on $\mathbb{R}^2$. Suppose that f is a contraction map on the metric space $\left(\mathbb{R}^2, d\right)$ with contraction factor e. Then the extension of f is a contraction on $(\mathcal{K}, h)$ with contraction factor e.*

Proof. Because one point sets $\{x\}$ are in $\mathcal{K}$, the contraction factor (if it exists) of the extended map is no smaller than e since

$$h(\{f(x)\}, \{f(y)\}) = d(f(x), f(y)) \leq ed\,(x, y) = eh(\{x\}, \{y\})$$

and e is the least such number for which the inequality holds. To show that e is the contraction factor of the extension, suppose that E and F are in $\mathcal{K}$ and let $a > h(E, F)$. For each x in E, there exists a corresponding y in F such that $d(x, y) < a$ which implies that $d(f(x), f(y)) < ea$. Thus, $f(E) \subseteq B(f(F), ae)$. Symmetrically, $f(F) \subseteq B(f(E), ae)$. Consequently, $h(f(E), f(F)) \leq eh(E, F)$. □

The concept of a contraction map on $\mathcal{K}$ is extended to finite sets of maps on $\mathcal{K}$ in the next section.

6.2 Iterated function systems

One of the more profound and exciting developments in the application of contraction maps is in the construction of fractal sets using iterated function systems. The mathematics was developed by John Hutchinson [Hutchinson (1981)] and the method was popularized by Michael Barnsley [Barnsley (1993)] and others. The iterated function system approach provides a good theoretical framework from which to pursue the mathematics of many classical fractals as well as more general types. Throughout we will use the complete metric space $\left(\mathbb{R}^2, d\right)$ and $\mathcal{K}$ will be the set of nonempty compact subsets of $\mathbb{R}^2$.

Definition 6.5. An **iterated function system** (IFS) is a finite collection $\{f_1, \ldots, f_m\}$ of contraction maps. A set $F \in \mathcal{K}$ is **invariant** for the maps f_i if

$$F = \bigcup_{i=1}^{m} f_i\,(F)\,.$$

Such invariant sets are often fractals.

The m mappings in an IFS are used to construct a single contraction mapping T on the space $\mathcal{K}$. The Hutchinson transformation $T : \mathcal{K} \to \mathcal{K}$ is defined by

$$T\,(E) = f_1\,(E) \cup f_2\,(E) \cup \cdots \cup f_m\,(E)\,,\, E \in \mathcal{K}.$$

The domain of the Hutchinson transformation consists of compact sets. Because continuous images of compact sets are compact and a finite union of compact sets is compact we also have that the points in the range are compact sets.

Our next result shows that the Hutchinson transformation is a contraction map on $(\mathcal{K}, h)$. Thus, the Hutchinson transformation has a unique fixed point called the **attractor** of the IFS.

Theorem 6.2. *Suppose that* $\{f_j\}_{j\leq n}$ *is a finite set of contraction maps on* $\mathcal{K}$, *with corresponding contraction factors* $\{e_j\}_{j\leq n}$. *Put* $e = \max\{e_j : j \leq n\}$. *For* $E \in \mathcal{K}$, *put*

$$T(E) = \bigcup_{j=1}^{n} f_j(E).$$

Then T *is a contraction map on* $\mathcal{K}$ *with contraction factor no larger than* e.

Proof. Let $E \in \mathcal{K}$ and $F \in \mathcal{K}$. We verify that $h(T(E), T(F)) \leq eh(E, F)$ by showing that $h(T(E), T(F)) \leq ea$ whenever $a > h(E, F)$. For $j \leq n$,

$$f_j(E) \subseteq B(f_j(F), e_j a) \subseteq B(T(F), e_j a) \subseteq B(T(F), ea).$$

Consequently,

$$T(E) = \cup_{j\leq n} f_j(E) \subseteq B(T(F), ea).$$

By symmetry, $T(F) \subseteq B(T(E), ea)$. Consequently, $h(T(E), T(F)) \leq ea$. □

Example 6.3. The standard Cantor set C can be described as an invariant set. Let $f_1, f_2 : \mathbb{R} \to \mathbb{R}$ be the similarities given by

$$f_1(x) = \frac{1}{3}x \text{ and } f_2(x) = \frac{1}{3}x + \frac{2}{3}.$$

Then $f_1(C)$ and $f_2(C)$ are just the left and right 'halves' of C, so that $C = f_1(C) \cup f_2(C)$. Thus C is invariant for the mappings f_1 and f_2. The collection $\{f_1, f_2\}$ is an IFS on $\mathbb{R}$.

Theorem 6.3. *Let* $\{f_1, f_2, \cdots, f_m\}$ *be an IFS on* $\mathcal{K}$. *Given any initial set* $E_0 \in \mathcal{K}$ *the sequence* $E_n = T(E_{n-1}), n = 1, 2, \cdots$, *where* T *is the Hutchinson transformation, converges in the Hausdorff metric to a unique attractor set* $E \in \mathcal{K}$. *Alternatively, the set* E *can be described as*

$$E = \lim_{n\to\infty} T^{(n)}(E_0).$$

The set E is a nonempty compact invariant set.

6.3 An iterated function system defines a curve

Throughout this section we restrict our attention to the metric space $(\mathbb{R}^2, d)$. Many of the maps in previous chapters correspond to iterated function systems. For example, the Koch curve is the fixed point (attractor) of the IFS defined by the contraction mappings

$$S_j(z) = C(j-1) + D(j-1)z, 1 \leq j \leq 4,$$

where the constants are given in Section 2.1 of Chapter 2. The continuous map from I to $\mathbb{R}^2$ whose image is the Koch curve is not included in the IFS approach.

What characteristics are required for the attractor of an IFS to be a curve? The following example demonstrates that the fixed set of an IFS can be disconnected with a non-trivial connected component and an isolated point.

Example 6.4. Let $f((x,y)) = (0,1)$, $g((x,y)) = (x/2, 0)$, and $h((x,y)) = 1/2 + (x/2, 0)$. Set $F(E) = f(E) \cup g(E) \cup h(E)$. The fixed set is $E = \{(0,1)\} \cup [0,1]$ in the plane.

We will use the Hahn-Mazurkiewicz Theorem ([Sagan (1994)], p.98) to show that if the attractor of an IFS is connected, then it is a curve.

Definition 6.6. A set X in $\mathbb{R}^n$ is **locally connected** if for every point $p \in X$ and every $\epsilon > 0$ there is an $\eta(p,\epsilon) > 0$ such that for every point $q \in B(p, \eta(p,\epsilon)) \cap X$ there is a compact and connected subset of X that contains p and q and is contained in $B(p, \epsilon)$.

The difficult half of the Hahn-Mazurkiewicz Theorem asserts that a connected compact set that is locally connected is a curve. The easier half is

Theorem 6.4. *A curve is locally connected.*

Proof. Let $p \in K$, where K is the image of a continuous map $f : [0,1] \to \mathbb{R}^n$. Put $S = f^{-1}(p) = \{y \in [0,1] : f(y) = p\}$. Let $0 < \epsilon < \text{dia}(K)/2$ and $\delta > 0$ be such that $|f(x) - f(y)| < \epsilon$ if $|x - y| \leq \delta$. Set $V = f([0,1]/B(S,\delta))$. Let $\eta = \eta(p,\epsilon) < d(p, V)$ and let $q \in B(p, \eta)$. Then $q = f(y)$ where $d(S,y) < \delta$. Choose $x \in S$ with $|x - y| = d(S,y)$. Thus $f([x,y])$ is a compact, connected subset of K that contains p and q and is contained in $B(x,\epsilon)$. □

Theorem 6.5. *Suppose the fixed set E*

$$E = \bigcup_{j=1}^{k} f_j(E)$$

of an IFS $\{f_1, \cdots, f_k\}$ is connected. Then E is a curve.

Proof. In view of the Hahn-Mazurkiewicz Theorem, it suffices to show that E is locally connected. To this end, let $p \in E$ and let $0 < \epsilon < \text{dia}(E)/2$. We will define an appropriate $\eta(\epsilon, p)$ below. Denote the contraction factor of f_j by e_j. Put $e = \max\{e_j : j \leq k\} < 1$. For $n \in \mathbb{N}$, E is composed of the union of connected compact sets

$$E_{j_1 \cdots j_n} = f_{j_n} \circ \cdots \circ f_{j_1}(E).$$

Each of these sets has diameter less than or equal to $d_n = e^n \text{dia}(E)$. Choose n such that $d_n < \epsilon$. Then p is in some, but not all, of the sets $E_{j_1 \cdots j_n}$. Put $\eta(\epsilon, p) = \min\{d(p, E_{j_1 \cdots j_n}) : p \notin E_{j_1 \cdots j_n}\}$. Fix a point

$$q \in B(p, \eta(\epsilon, p)) \cap E.$$

Then, for some $E_{j_1 \cdots j_n}$,

$$p, q \in E_{j_1 \cdots j_n} \subset B(p, \epsilon).$$

□

According to Theorem 6.5 the fixed point E of an IFS $\{f_1, \cdots, f_k\}$ is a curve if and only if E is connected. A necessary and sufficient condition that depends only on deterministic iterates of the Hutchinton transformation follows.

Theorem 6.6. *The attractor E of an IFS $\{f_1, \cdots, f_k\}$ is connected if and only if there is a compact set F such that the sequence $\{F_n\}_{n \geq 1}$, where*

$$F_1 = T(F) = \bigcup_{j=1}^{k} f_j(F)$$

and $F_{n+1} = T(F_n)$, is composed of connected sets.

Proof. If the fixed point E is connected, then put $F = E$. Problem C.27 and the fact that $F_n \to E$ in the Hausdorff metric imply that the stated conditions are sufficient. □

Example 3.8 is an application of Theorem 6.6. Another application follows.

Example 6.5. Define the linear map $L : \mathbb{R}^2 \to \mathbb{R}^2$ by

$$L = \begin{bmatrix} 1 & -1 \\ 1 & 1 \end{bmatrix};$$

then

$$M = L^{-1} = \begin{bmatrix} 1/2 & 1/2 \\ -1/2 & 1/2 \end{bmatrix} = \frac{1}{\sqrt{2}} M_{-\pi/4}.$$

The two maps $\{f_1, f_2\}$ with $f_1 = M$ and $f_2 = (0,1) + M$ comprise an IFS with the Hutchinson transformation

$$T(E) = f_1(E) \cup f_2(E).$$

Let U be the unit square. Then $T(U)$ is the union of two smaller squares arranged as shown in Figure 6.2a. Note that the corners of U remain in $T(U)$ and $(1/2, 1/2) \in f_1(U) \cap f_2(U)$. Figure 6.2b demonstrates that the corners of U remain in $f_1(T(U)) \cup f_2(T(U))$ and $(1/2, 1/2) \in f_1(T(U)) \cap f_2(T(U))$. Similarly, you can verify that $(1/2, 1/2) \in f_1(T^n(U)) \cap f_2(T^n(U))$, for $n \in \mathbb{N}$. The fixed point E of this IFS is a curve often called the Twin Dragon. A representation of E, with the corners of U and the point $(1/2, 1/2)$ marked, is shown in Figure 6.2c.

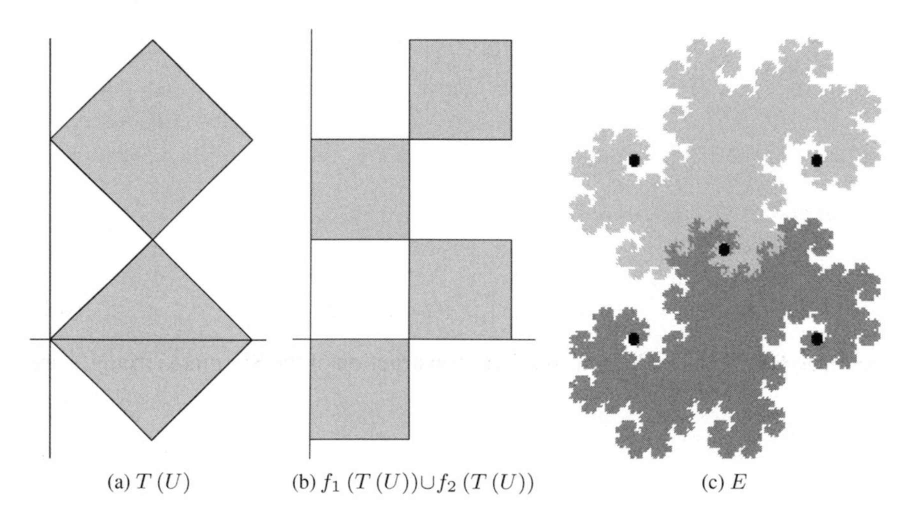

(a) $T(U)$ (b) $f_1(T(U)) \cup f_2(T(U))$ (c) E

Fig. 6.2: Construction of the Twin Dragon.

6.4 Implementation of iterated function systems

There are two approaches to implementing an IFS: a deterministic approach and a random approach.

Algorithm 6.1 (The Deterministic Algorithm). *Let $\{f_j : j = 1, \cdots, n\}$ be an IFS. Choose a compact set $E_0 \subset \mathbb{R}^2$. Then compute successively $E_{k+1} =$*

$\cup_{j=1}^{n} f_j(E_k)$*, for* $k = 0, 1, \cdots$*. By the previous section, the sequence* $\{E_k\}$ *converges to the attractor of the IFS in the Hausdorff metric.*

We used the deterministic algorithm in our construction of the Koch curve in Chapter 2. The iterated function system $\{S_j, j = 1, \ldots, 4\}$ is defined by

$$S_j(z) = C(j-1) + D(j-1)z.$$

Beginning with $E_0 = I$, then each of the curves $E_{k+1} = f_{j+1}(I) = S_1(E_k) \cup S_2(E_k) \cup S_3(E_k) \cup S_4(E_k)$ is a compact set in $\mathbb{R}^2$. The sequence $\{f_j(I)\}$ converges in the Hausdorff metric to the Koch curve.

Example 6.5 also demonstrates the deterministic algorithm, using matrices to define the IFS on $\mathbb{R}^2$. Problem 6.3 illustrates that a matrix M can have two eigenvalues with absolute value less than one and yet not be a contraction map. However, the exercise also shows that a power of M is a contraction map. In this section we assume that all matrices M satisfy the condition that a power of M is a contraction map. We demonstrate that such matrices can be used to generate fractal images.

Let n denote a positive integer and, for each $j = 1, \cdots, n$, let

$$M_j = \begin{bmatrix} a_j & c_j \\ b_j & d_j \end{bmatrix}$$

and $t_j = \begin{bmatrix} u_j & v_j \end{bmatrix}^T \in \mathbb{R}^2$. We restrict attention to the IFS on the metric space $\mathbb{R}^2$ using contractions

$$f_j(z) = M_j z + t_j, z \in \mathbb{R}^2. \tag{6.1}$$

Example 6.6. Steps in the deterministic construction of the Sierpinski triangle are shown in Figure 1.10. There E_0 is chosen as the equilateral triangle and the IFS is the set of functions $\{f_j\}$, with

$$M_j = \begin{bmatrix} 1/2 & 0 \\ 0 & 1/2 \end{bmatrix}$$

and

$$t_1 = (0,0), t_2 = (1/2, 0), t_3 = \left(1/4, \sqrt{3}/4\right).$$

Example 6.7. Choose $M = \begin{bmatrix} 1/4 & 0 \\ 0 & 1/4 \end{bmatrix}$ and the original compact set E_0 as the unit square. The sixteen maps of the IFS will be of the form $f_j(z) = Mz + t_j$, $j = 1, \cdots, 16$, where the t_j are given in Table 6.1. Figure 6.3a shows E_1, Figure 6.3b shows E_2, and Figure 6.3c shows E_5. The sequence $\{E_j\}$ converges in the Hausdorff metric to the Badge and Hydrant curve shown in Figure 1.12 [Drenning *et al.* (2005)].

Table 6.1: Values of t_j for the Badge and Hydrant transformations.

$(0,0)$	$(0,3/4)$	$(3/4,0)$	$(3/4,3/4)$
$(-1/4,1/4)$	$(-1/4,1/2)$	$(1/4,-1/4)$	$(1/4,1/4)$
$(1/4,1/2)$	$(1/4,1)$	$(1/2,-1/4)$	$(1/2,1/4)$
$(1/2,1/2)$	$(1/2,1)$	$(1,1/4)$	$(1,1/2)$

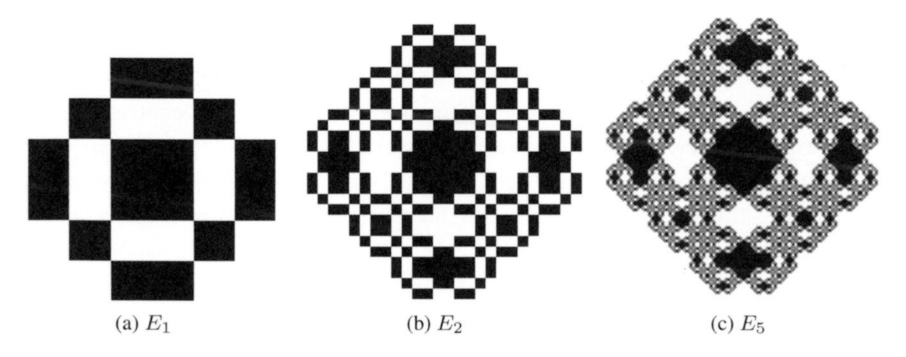

(a) E_1 (b) E_2 (c) E_5

Fig. 6.3: Three stages of the Badge and Hydrant curve.

Table 6.2: Values of t_j for Example 6.8.

$\left(0,\sqrt{3}\right)$	$\left(1,\sqrt{3}\right)$	$\left(2,\sqrt{3}\right)$
$\left(1/2,\sqrt{3}/2\right)$	$\left(1/2,3\sqrt{3}/2\right)$	$\left(5/2,3\sqrt{3}/2\right)$
$\left(0,2\sqrt{3}\right)$	$\left(1,2\sqrt{3}\right)$	$\left(3/2,3\sqrt{3}/2\right)$

Example 6.8. Again we allow E_0 to be an equilateral triangle with lower left corner at the origin. Define 9 maps

$$f_j(z) = \begin{bmatrix} 1/3 & 0 \\ 0 & 1/3 \end{bmatrix}(z) + t_j,\ j = 1,\ldots,9$$

where the t_j are given in Table 6.2. The sets E_1 and E_2 are shown in Figure 6.4 along with the limiting curve.

Example 6.9. This example is similar to the previous one except for the selection of the nine triangles. You are asked to define the nine maps that will produce the snowflake curve shown Figure 6.5b. (See Problem 6.16.)

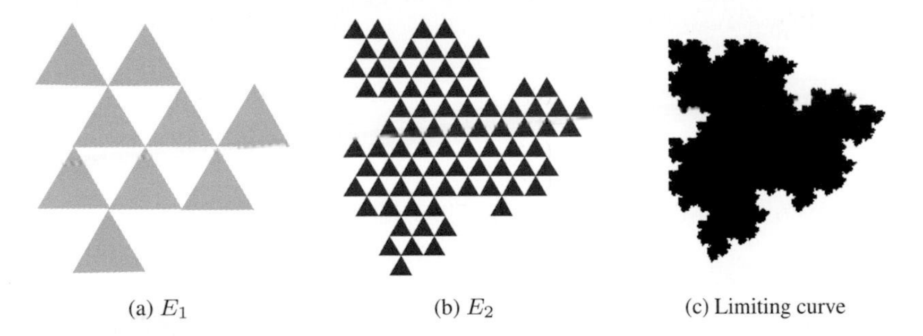

(a) E_1 (b) E_2 (c) Limiting curve

Fig. 6.4: First two steps and final figure for Example 6.8.

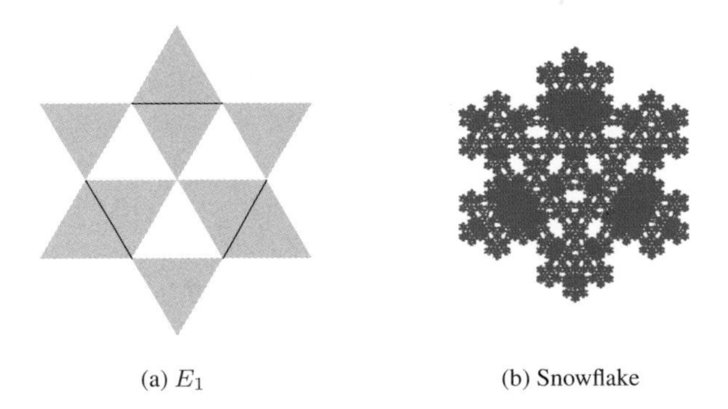

(a) E_1 (b) Snowflake

Fig. 6.5: First step limit snowflake curve. (See Example 6.9)

The random approach is different from the deterministic in that we begin with a single arbitrary point in $\mathbb{R}^2$. At each subsequent step a new point is calculated by evaluating the previous point using a randomly chosen member of the IFS. After the first few, these points "visually belong" to the attractor.

Algorithm 6.2 (The Random Iteration Algorithm). *Let* $\{f_j : j = 1, \cdots, n\}$ *be an IFS. Choose* $z_0 \in \mathbb{R}^2$ *and then choose recursively, independently,*

$$z_k \in \{f_1(z_{k-1}), f_2(z_{k-1}), \cdots, f_n(z_{k-1})\}$$

for $k = 1, 2, \cdots$. *Next, throw away the first* N *(say* $N = 1000$*) iterates and use the set* $\{z_N, \ldots, z_n\}$ *as an approximation to the attractor. For large* n *(say* $n = 50000$*), this set provides a visual representation of the attractor. (See [Barnsley (1993)].)*

Most of the fractal images in this book are generated using the Random Iteration Algorithm.

6.5 Problems

PROBLEM 6.1. Prove that a contraction mapping is uniformly continuous.

PROBLEM 6.2. Let $A = A\left(\mathbb{R}^2\right)$ and $M \in A$. Suppose M has two real eigenvalues with absolute value less than 1. Show that $\lim_{k\to\infty} \left\|M^{(k)}\right\| = 0$.

PROBLEM 6.3. Recall Problem 5.34 where it was shown that the matrix $U = \begin{bmatrix} 1.2 & 1.1 \\ 0 & 1.1 \end{bmatrix}$ has eigenvalue $\lambda = 1.1$, and, if $M = U^{-1}$, then $\|M\| > 1$. Now verify that M is a contraction map on the metric space $\left(\mathbb{R}_2, \|\cdot\|_\phi\right)$ with contraction factor $e = 1/1.1 = \lambda^{-1}$. Show that there is a positive integer k for which $\left\|M^{(k)}\right\| < 1$.

PROBLEM 6.4. Complete the proof of Theorem 6.6.

PROBLEM 6.5. Define contraction mappings that will generate a right Sierpinski triangle.

PROBLEM 6.6. Define 5 contraction mappings that will generate a self-similar pentagon as shown in Figure 6.6.

PROBLEM 6.7. Define 6 contraction mappings that will generate a self-similar hexagon as shown in Figure 6.7.

PROBLEM 6.8. Show that the one point subsets of $\mathbb{C}$ are in $\mathcal{K}$.

PROBLEM 6.9. Let f be a continuous map from $\mathbb{C}$ to $\mathbb{C}$. Show that the extension of f is a continuous map on $\mathcal{K}$.

PROBLEM 6.10. Suppose that f is a contraction map on $\mathbb{C}$ with fixed point t. Find the fixed point of the extension of f to $\mathcal{K}$.

PROBLEM 6.11. Suppose that $\{E_j\}$ is a sequence of compact subsets of $\mathbb{C}$.

(a) Let n be a positive integer. Show that $\cup_{j\leq n} E_j$ is a compact subset of $\mathbb{C}$.

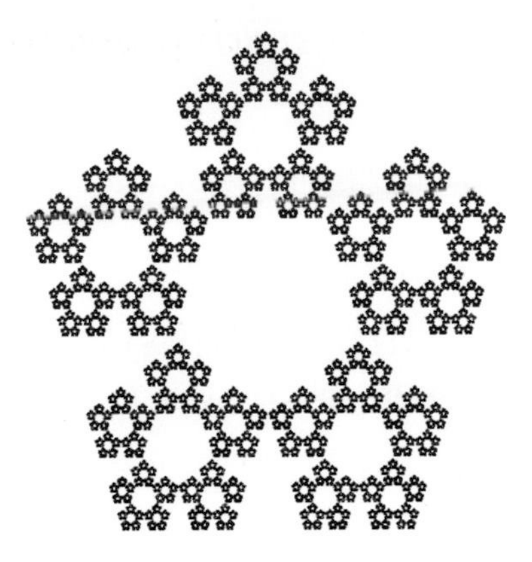

Fig. 6.6: Pentagon.

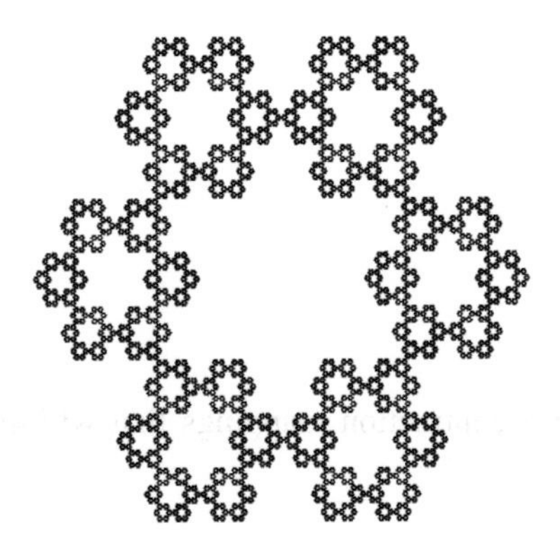

Fig. 6.7: Hexagon.

(b) Show that $\cap_{j \geq 1} E_j$ is a compact subset of $\mathbb{C}$.

PROBLEM 6.12. Let $f_1(x) = (1/2)x$, let $f_2(x) = 1/2 + (1/2)x$, and let $f(E) = f_1(E) \cup f_2(E)$. Find the fixed point in $\mathcal{K}$ of f.

PROBLEM 6.13. Let $f_1(x) = (1/3)x$, let $f_2(x) = 2/3 + (1/3)x$, and let $f(E) = f_1(E) \cup f_2(E)$. Find the fixed point in $\mathcal{K}$ of f.

PROBLEM 6.14. Let $f_1(z) = (1/3)z$, let $f_2(z) = 2/3 + (1/3)z$, let $f_3(z) =$

$4/3 + (1/3)z$, and let $f(E) = f_1(E) \cup f_2(E) \cup f_3(E)$. Find the fixed point in $\mathcal{K}$ of f.

PROBLEM 6.15. Let $f_1(z) = (1/3)z$, let $f_2(z) = 2/3 + (1/3)z$, let

$$f_3((x, y)) = (1/2, \sqrt{3}/2) + \frac{e^{-i\pi/2}}{3} z,$$

and let $f(E) = f_1(E) \cup f_2(E) \cup f_3(E)$. Find the fixed point in $\mathcal{K}$ of f. Hint: Let E be the equilateral Sierpinski triangle; compute $f^{(1)}(E)$ and $f^{(2)}(E)$.

PROBLEM 6.16. Determine the nine maps that will produce the snowflake curve in Figure 6.5b as described in Example 6.9.

PROBLEM 6.17. Show that if the fixed set of an IFS is self-similar and disconnected, then it is homeomorphic to the standard Cantor set.

Chapter 7

Dimension, curves and Cantor sets

You are familiar with objects of elementary geometry with dimension $1, 2$ or 3. A line interval is one dimensional; it has length. A square is two dimensional; it has area. A cube is three dimensional; it has volume. In this chapter we indicate how to extend the concept of dimension to all bounded sets in $\mathbb{R}$ and $\mathbb{R}^2$. Establishing the dimension of a set is often a two step process. First we obtain a good estimate for the dimension of the set; then we need to show that the estimate is accurate.

Mathematicians have given various technical definitions for the dimension of sets. We use Hausdorff dimension. We motivate it by first considering intervals, squares, and cubes and develop a process that leads to corresponding values 1,2, and 3 for their dimensions. Following these examples, we give the definition of Hausdorff dimension. Next, we look at a family of Cantor sets that have noninteger dimension. For any d, $0 < d < 1$, we show that a member of this family has dimension d. Since we know that these sets evolve via repetition of a simple geometric process, we use the process to estimate the dimension of the set and then assert that the estimate is the dimension of the set. The mathematical details of showing that an estimate is correct are beyond the scope of this book. Similarly, for any d, $1 < d < 2$, we use a member of this Cantor set family to describe a simple curve with dimension d. Compact sets with positive area have Hausdorff dimension equal to 2. Hausdorff dimension enables us to compare sizes of sets with area equal to zero; this comparison is consistent with the geometry of the tent maps that generate the Cantor sets.

7.1 Intervals, squares and cubes

What do we know about intervals, squares, and cubes? An interval has zero area and volume. A square has infinite length, since we can put very long simple curves in a square, but it has zero volume. A cube has infinite area and length since we can put infinitely many pairwise disjoint squares in a cube.

We can cover the unit interval $I = [0, 1]$ with n intervals of length $1/n$. We can also cover I with n squares of side $1/n$ or n cubes of side $1/n$. We cannot cover the unit square $U = I \times I$ with finitely many intervals; however, we can cover it with n^2 squares of side $1/n$ or n^2 cubes of side $1/n$. Similarly, while we cannot cover the unit cube $I^3 = I \times I \times I$ with finitely many lines or squares, we can cover I^3 with n^3 cubes of side $1/n$. Based on what we know, we define and examine limits of three functions. Then we use these limits to estimate the dimensions of an interval, a square, and a cube.

For the unit interval, consider the function L_d defined for $n \in \mathbb{N}$ by the formula

$$L_d(n) = n(1/n)^d = n^{1-d}.$$

If $d < 1$, then $\lim_n L_d(n) = \infty$; if $d = 1$, then $L_d(n) = 1$ for $n \geq 1$; if $d > 1$, then $\lim_n L_d(n) = 0$. For squares, suppose the function A_d is defined for $n \in \mathbb{N}$ by the formula

$$A_d(n) = n^2(1/n)^d = n^{2-d}.$$

If $d < 2$, then $\lim_n A_d(n) = \infty$; if $d = 2$, then $A_d(n) = 1$; if $d > 2$, then $\lim_n A_d(n) = 0$. For cubes, let the function V_d be defined, for $n \in \mathbb{N}$, by the formula

$$V_d(n) = n^3(1/n)^d = n^{3-d}.$$

If $d < 3$, then $\lim_n V(n) = \infty$; if $d = 3$, then $V_d(n) = 1$; if $d > 3$, then $\lim_n V_d(n) = 0$. The **critical value** for d, where the limit changes from ∞ to 0, is 1 for an interval, 2 for a square, and 3 for a cube. According to this discussion, we estimate the dimension of an interval, a square, and a cube to be equal to 1, 2, and 3, respectively. By the definition of dimension given below, our estimates are upper bounds for the dimensions of these sets.

More generally, if we have an interval $[a, b]$ then n intervals of length $\dfrac{b-a}{n}$ cover it. Also, $n\left(\dfrac{(b-a)}{n}\right)^d = (b-a)^d n^{1-d}$ and number $(b-a)$ does not change the critical value of d. Intervals may have different lengths, but they all have dimension 1. Later we demonstrate that corresponding comments apply to squares, cubes, and other sets.

7.2 Hausdorff dimension of a bounded subset of $\mathbb{R}^2$

A countable set $\{K_j\}_{j\geq 1}$ of closed squares is said to be a **countable cover** of a bounded set $S \subset \mathbb{R}^2$ if S is a subset of the union of the squares:

$$S \subseteq \bigcup_{j\geq 1} K_j.$$

The set $\{K_j\}_{1\leq j}$ of closed squares is said to have **size** at most ϵ if each has side length at most ϵ. Let U_m denote the set of countable covers of S with size at most $1/m$. Let $d > 0$ and set

$$h_m(S,d) = \inf\left\{\sum_{1\leq j} \text{sidelength}\,(K_j)^d : \{K_j\}_{1\leq j} \in U_m\right\}.$$

For $m \geq 1$,

$$h_m(S,d) \leq h_{m+1}(S,d).$$

Set

$$h(S,d) = \lim_{m\to\infty} h_m(S,d).$$

Definition 7.1. The **Hausdorff dimension** $H(S)$ of S is

$$H(S) = \inf\{d : h(S,d) = 0\}.$$

A countable cover is either finite or countably infinite. (Infinite sets are discussed in Appendix D.) Finite covers suffice to define the Hausdorff dimension of a compact set.

7.2.1 *Basic facts about dimension*

- **Fact 1:** Dimension depends on the shape of a set, not on its size. If S is a set in $\mathbb{R}^2$, c is a positive number, and $cS = \{cs : s \in S\}$, then S and cS have the same dimension.
- **Fact 2:** If S and T are subsets of $\mathbb{R}^2$ and $S \subseteq T$, then the dimension of S is less than or equal to the dimension of T.
- **Fact 3:** If S_j and T are subsets of $\mathbb{R}^2$ with $H(S_j) \leq H(T)$, $j \in \mathbb{N}$, then

$$H(T \cup \bigcup_{j\geq 1} S_j) = H(T).$$

- **Fact 4:** If S is a subset of $\mathbb{R}$, and $\text{len}(S) > 0$, then $H(S) = 1$.

- **Fact 5:** If S is a curve, then $H(S) \geq 1$.
- **Fact 6:** If S is a subset of $\mathbb{R}^2$, and $\text{area}(S) > 0$, then $H(S) = 2$.
- **Fact 7:** If S is a countable set, then $H(S) = 0$.

Example 7.1. We can cover the standard Cantor set C with 2^n intervals, squares, or cubes of side $1/3^n$. Then

$$2^n(1/3^n)^d = 1$$

if and only if

$$n\ln(2) + nd\ln(1/3) = n\ln(2) - nd\ln(3) = 0$$

if and only if

$$d = \frac{\ln(2)}{\ln(3)}.$$

If $d < \ln(2)/\ln(3)$, then $\lim_{n\to\infty} 2^n(1/3^n)^d = \infty$. If $d > \ln(2)/\ln(3)$, then $\lim_{n\to\infty} 2^n(1/3^n)^d = 0$. Therefore, we estimate the dimension of the standard Cantor set to be equal to $\ln(2)/\ln(3)$. We do not explain why $H(C) = \ln(2)/\ln(3)$.

7.3 Tent maps and Cantor sets with prescribed dimension

As we have seen in Chapter 4, the standard Cantor set $C = \widetilde{C}_{1/3}$ is one of a family of Cantor sets of length zero. These Cantor sets also emerge from iterates of tent maps g_h as described below. A generic tent map is defined by

$$g_h(x) = 2h(1/2 - |1/2 - x|)$$

where $h > 1$. An example is shown in Figure 7.1.

Starting with the initiator $g_h^{(0)}(x) = x$, set

$$g_h^{(n+1)}(x) = g_h(g_h^{(n)}(x)),\ n \geq 1.$$

Fix $x_0 \in I$. The sequence $\{x_n = g_h^{(n)}(x_0)\}_{n\geq 0}$, is called the **orbit** of x_0. Either the orbit of x_0 stays in I as demonstrated in Example 7.2 or, as demonstrated in Example 7.3, $\lim_{n\to\infty} g_h^{(n)}(x_0) = -\infty$. To see this suppose $\left|g_h^{(n)}(x_0)\right| > 1$ for some $n \geq 1$. Then

$$-\left|g_h^{(n)}(x_0)\right| > g_h^{(n+1)}(x_0) > \cdots \longrightarrow -\infty.$$

Example 7.2. A graphical way to trace the orbit of $x_0 = 0.3$ determined by $g_{3/2}(x) = 3(1/2 - |1/2 - x|)$ is shown in Figure 7.2. First evaluate $x_1 =$

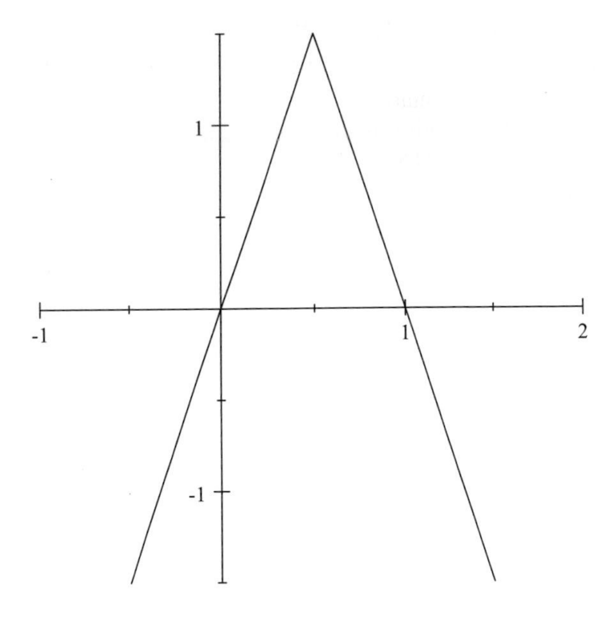

Fig. 7.1: Generic tent with $h = 3/2$.

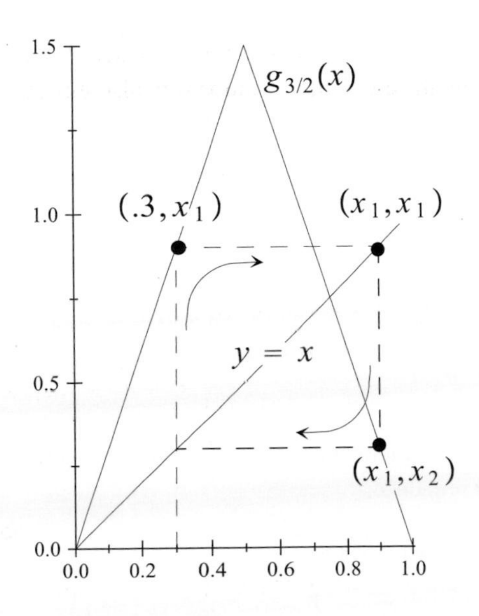

Fig. 7.2: Orbit map for $x_0 = 0.3$ using $g_{3/2}$.

$g_{3/2}(x_0) = 0.9$; then locate $(0.9, 0.9)$ on the line $y = x$. Trace vertically from that point to $(0.9, 0.3)$ where $0.3 = g_{3/2}(0.9)$. In this example $x_2 = x_0$ and the sequence repeats producing the orbit $\{0.3, 0.9, 0.3, 0.9, \cdots\}$. Similarly, the orbit of $x_0 = 0$ is just $\{0\}$; the orbit of $x_0 = 1$ is $\{0, 1\}$; the orbit of $x_0 = 1/3$ is $\{0, 1/3, 1\}$; and the orbit of $x_0 = 1/4$ is $\{1/4, 3/4\}$.

The next example demonstrates an orbit that tends to $-\infty$.

Example 7.3. Using $x_0 = 0.4$ the tent map $g_{3/2}$ produces the orbit $\{0.4, 6/5, -3/5, -9/5, -27/5, \ldots\}$ as suggested by Figure 7.3.

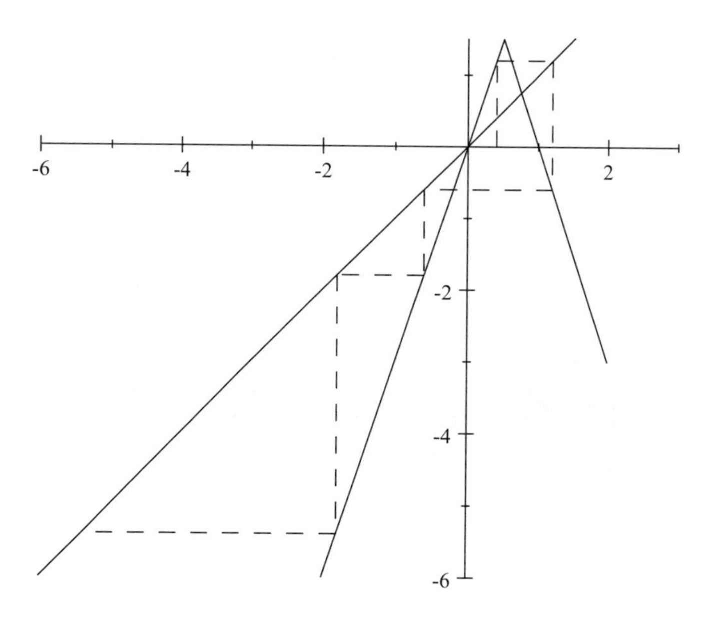

Fig. 7.3: Example of a divergent orbit.

Definition 7.2. The **resident set** S_h for g_h is the set of points x for which the orbit of x is a bounded sequence.

For a fixed $h > 1$ set

$$G_n = \left\{ x : 0 \leq g_h^{(n)}(x) \leq 1 \right\}.$$

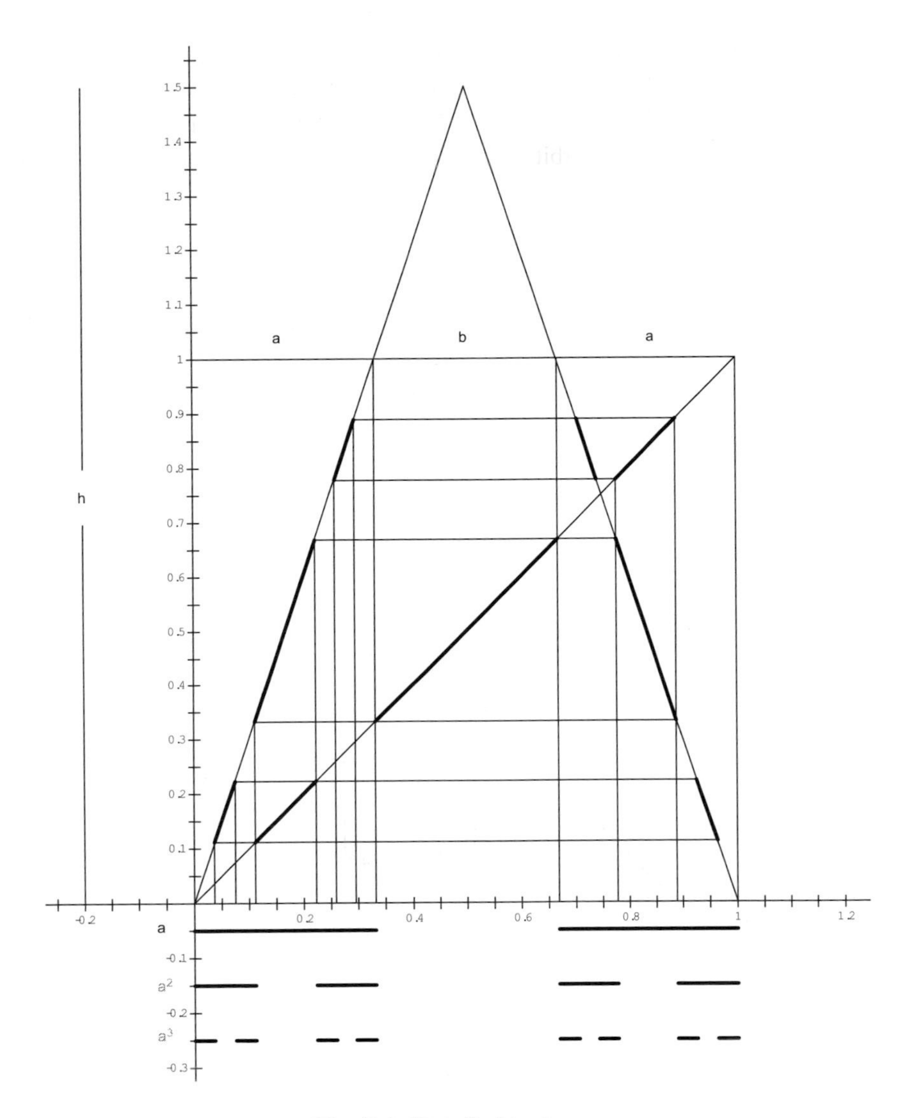

Fig. 7.4: Detailed tent map.

To describe G_n, set $a = 1/(2h) < 1/2$. Then G_n is composed of 2^n intervals of length a^n as shown in Figure 7.4. The Cantor-like behavior is suggested by the darkened horizontal line segments below the graph.

The dark line segments on the graph show the image of points in $I \backslash G_n$ for $n = 1, 2, 3$. Recall that $g_h^{(0)}(x) = x$. Then $I = G_0 \supset G_1 \supset G_2 \cdots$.

7.3.1 *Dimension of Cantor sets*

For $0 < a < 1/2$, recall that

$$\widetilde{C}_a = \bigcap_{n \geq 0} G_n$$

denotes the Cantor set generated in $[0, 1]$ by the sequence $\{G_n\}$. (The Cantor sets $\widetilde{C}_a$ were defined in Subsection 4.3.2 of Chapter 4.) The taller the tent, the smaller the Cantor set that is generated. However, all of these Cantor sets have length equal to zero. Consequently, length does not distinguish their relative sizes.

The set $\widetilde{C}_a$ can be covered by 2^n intervals of length a^n. The equation for estimating the dimension of $\widetilde{C}_a$ is

$$2^n \left(a^n\right)^d = 1.$$

Consequently,

$$d = \frac{\ln(2)}{\ln(1/a)}.$$

If $d < \ln(2)/\ln(1/a)$, then $\lim_n 2^n \left(a^n\right)^d = \infty$; if $d > \ln(2)/\ln(1/a)$, then $\lim_n 2^n \left(a^n\right)^d = 0$. As a goes from 0 to $1/2$, $1/a$ goes from infinity to 2, and the critical value $d = \ln(2)/\ln(1/a)$ goes from 0 to 1. We can see that the critical value of d distinguishes the sizes of the Cantor sets $\widetilde{C}_a$ in a reasonable way. This critical value can be verified as the Hausdorff dimension of the Cantor set $\widetilde{C}_a$. With $0 < d < 1$, if we set $a = 1/2^{1/d}$, we see that $H(\widetilde{C}_a) = d$.

We saw in Chapter 4 that the family $\{\widetilde{C}_a\}_{0<a<1/2}$ of Cantor sets arises from generalized Koch curves. Note that we have $d = 1/2$ when $a = 1/4$, the value used to generate the RS curve. Also $d > 1/2$ when $a > 1/4$; that is, $d > 1/2$ if and only if the corresponding generalized Koch curve is a simple curve.

Example 7.4. We display a set S in I with length equal to 0 and dimension equal to 1. Let $\{a_k\}$ be a sequence of positive numbers strictly increasing to $1/2$. The dimension of $\widetilde{C}_{a_k}$ is

$$d_k = \frac{\ln(2)}{\ln(1/a_k)} \nearrow 1.$$

We shrink $\widetilde{C}_{a_k}$ to length $1/2^k$, translate the shrunken set to fit in the interval $[1/2^k, 1/2^{k-1}]$ and label the translated set S_k. Set

$$S = \{0\} \cup \bigcup_{k \geq 1} S_k.$$

Since $S_k \subset S$, the dimension of S is $\geq d_k$ for $k \geq 1$. Thus, the dimension of S is equal to 1. The length of S is equal to 0.

7.3.2 *Dimension of Cantor sets in the plane*

Next, we consider Cantor sets of the form $\widetilde{C}_a \times \widetilde{C}_a$, with area equal to zero in the plane. The corresponding dimension estimate is given by

$$4^n (a^n)^d = 1.$$

Consequently,

$$d = \frac{\ln(4)}{\ln(1/a)} = \frac{2\ln(2)}{\ln(1/a)}.$$

If $d < 2\ln(2)/\ln(1/a)$, then $\lim_n 4^n (a^n)^d = \infty$; if $d > 2\ln(2)/\ln(1/a)$, then $\lim_n 4^n (a^n)^d = 0$. As a goes from 0 to $1/2$, $1/a$ goes from infinity to 2, and the critical value $d = 2\ln(2)/\ln(1/a)$ goes from 0 to 2. As before, we call the critical value $d = 2\ln(2)/\ln(1/a)$ the dimension of the Cantor set $\widetilde{C}_a \times \widetilde{C}_a$. Note that as a goes from $1/4$ to $1/2$, d goes from 1 to 2.

7.4 Dimension and simple curves

7.4.1 *Simple curves with prescribed dimension*

The dimension of a curve is at least 1 and at most 2. In Section 3.2 we constructed a homeomorphism from C to $C \times C$, and we used variations of this construction to define simple curves in U with prescribed area y, where $0 < y < 1$. Curves with positive area have dimension 2. Here we use similar homeomorphisms from C to $\widetilde{C}_a \times \widetilde{C}_a$ to define simple curves $m(I)$ with prescribed dimension y, where $1 < y < 2$. Fix $y \in (1, 2)$, and choose $a \in (1/4, 1/2)$ such that

$$y = H(\widetilde{C}_a \times \widetilde{C}_a) = \frac{2\ln(2)}{\ln(1/a)}$$

as described in the previous section. Let m be the linear extension to I of the similar homeomorphism from C to $\widetilde{C}_a \times \widetilde{C}_a$. Because each of the countably many intervals that are added to the graph by the linear extension has Hausdorff dimension 1, we can appeal to basic Fact 3 and claim $H(m(I)) = H(m(C)) = y$. In Problem 7.4 you are asked to construct a simple curve with area 0 and dimension 2.

7.4.2 *Dimension of the Koch curve*

If we use squares or cubes to cover the Koch curve, there may be overlapping, which is permitted. However, we will cover the Koch curve with equilateral triangles that abut. These triangles are identified with squares in a way that should be clear. You can see in Figure 7.5 that 4^n triangles of side $1/3^n$ comprise an economical cover. Consequently, we use the equation $4^n(1/3^n)^d = 1$ to estimate the dimension of the Koch curve to be $d = 2\ln(2)/\ln(3)$. This estimate is, in fact, correct.

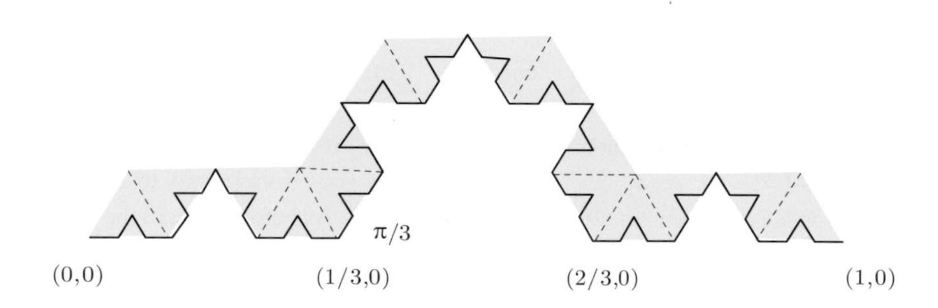

Fig. 7.5: Koch covered by triangles.

The Koch curve and the Cantor set $C \times C$ look very different; however, we see that the dimension of the Koch curve is equal to the dimension of $C \times C$. The dimensions we have assigned to sets in this chapter are upper bounds to their Hausdorff dimensions. These upper bounds are, in fact, equal to the Hausdorff dimensions of the sets, but we will not present the mathematics necessary to establish equality.

7.4.3 *Functions with prescribed dimension of points of non-tangency*

Fix $y \in (0,1)$. We will display a function $m : I \to I$ such that the set S of points x for which the graph of m has no tangent line at the point $(x, m(x))$ has dimension y; that is, $H(S) = y$. Each of the Cantor sets in the families $\{C_h\}_{0 \leq h < 1}$ and $\{\widetilde{C}_a\}_{0<a<1/2}$ has a corresponding Cantor function. Properties of tangent lines to the graphs of these curves have been studied in [Darst (1993)], [Darst (1995)], [Eidswick (1974)], and the references contained in these papers. We present a few pertinent properties of the graph G of the Cantor function corresponding to $T = \widetilde{C}_a$, although the mathematical details are too complicated to include.

The graph G has a horizontal tangent line at every point with x-coordinate in $I \backslash T$. It has no tangent line at a point when the x-coordinate of the point is an endpoint of T. If G has a tangent line at a point of G, then the tangent line is a vertical line.

The set S of points in I corresponding to x-coordinates of points in G with no tangent line at the point satisfies the equation

$$H(S) = \left(\frac{2\ln(2)}{\ln(1/a)}\right)^2.$$

Note that

$$0 < H(S) = \left(\frac{2\ln(2)}{\ln(1/a)}\right)^2 < H(T) = \frac{2\ln(2)}{\ln(1/a)}.$$

These inequalities imply that S is uncountable, but significantly more concentrated than T.

For a given $y \in (0, 1)$, choose a such that

$$\left(\frac{2\ln(2)}{\ln(1/a)}\right)^2 = y$$

and let m denote the Cantor function corresponding to $\widetilde{C}_a$. Then the set S of points x for which the graph of m has no tangent line at the point $(x, m(x))$ has dimension y; that is, $H(S) = y$.

7.5 Symmetric Cantor sets (optional section)

For $0 < d < 1$, we have displayed a corresponding Cantor set $\widetilde{C}_a$ in $[0, 1]$ with length 0 and dimension d. All the Cantor sets $\widetilde{C}_a$ are *symmetric* as defined below. After defining symmetric Cantor sets, we will display two symmetric Cantor sets in $[0, 1]$, each with length equal to 0. One has dimension equal to 0 and the other has dimension 1. (See Examples 7.5 and 7.6.)

7.5.1 *Construction of a symmetric Cantor set*

We outline an algorithm for constructing symmetric Cantor sets.

Step 1 Start with $I = [0, 1]$. Remove a segment $(a_1, 1 - a_1)$ of length $l_1 = 1 - 2a_1 < 1$ from the center of I, leaving 2 intervals of length a_1 at the ends of I. Denote the union of these 2 intervals by S_1.

Step 2 Remove a segment of length $l_2 < a_1$ from the center of each of the 2 intervals of length a_1, leaving 2^2 intervals of length a_2. Denote the union of these 2^2 intervals by S_2.

Continue recursively:

Step n+1 Remove a segment of length $l_{n+1} < a_n$ from the center of each of the 2^n intervals of length a_n, leaving 2^{n+1} intervals of length a_{n+1}. Denote the union of these 2^{n+1} intervals by S_{n+1} and let

$$S = \bigcap_{n \geq 1} S_n.$$

The set S is called a symmetric Cantor set. The length of S_n is $2^n a_n$, which converges monotonically to the length of S. The Cantor sets C_h from Chapter 1 are examples of the process. You should verify, using an appropriate function f_d, that they each have critical value $d = 1$ if $h > 0$. (See Problem 7.2.)

7.5.2 *Definition of dimension of symmetric Cantor sets*

We recall the dimension function and solve the equation

$$2^n (a_n)^{d_n} = 1$$

for d_n:

$$d_n = \frac{n \ln(2)}{\ln(1/a_n)}.$$

If the sequence $\{d_n\}$ converges to d, then d is defined to be the **dimension** of S.

Example 7.5. Set

$$a_n = 1/(n+1)^n.$$

Then

$$d_n = \frac{n \ln(2)}{\ln(1/a_n)} = \frac{\ln(2)}{\ln(n+1)} \to 0.$$

Example 7.6. Set

$$a_n \frac{1}{(n+1)2^n}$$

and solve the equation

$$2^n \left(\frac{1}{(n+1)2^n} \right)^{d_n} = 1$$

for d_n:

$$2^{n(1-d_n)} (n+1)^{-d_n} = 1 \Leftrightarrow$$

$$(1 - d_n) \ln(2) - d_n \frac{\ln(n+1)}{n} = 0;$$

$$d_n = \frac{\ln(2)}{\ln(2) + \frac{\ln(n+1)}{n}} \to 1.$$

In summary, we define the dimension d of a general symmetric Cantor set as

$$d = \lim_{k \to \infty} \left(\inf_{n \geq k} (d_n) \right),$$

where $2^n (a_n)^{d_n} = 1$.

7.6 Saw tooth maps

We conclude this chapter with a challenging project for you that remains an open mathematical question.

Project. Saw Tooth Maps

A tent map has its tip at the point $(1/2, h)$. We move the x coordinate of the tip to the right and consider a saw tooth map s with tip at (c, h), where $1/2 < c < 1$ and $h > 1$. The invariant set for s is homeomorphic to the Cantor set. Because $1/2 < c < 1$, the structure of the invariant set is more complicated. Reversals generate pieces of the invariant set that differ in length. (The corresponding pieces in a tent map's invariant set have equal lengths because tent maps are symmetric with respect to the vertical line $x = 1/2$.) The invariant set for a saw tooth map has a uniform distribution function. We leave properties of the invariant set, the distribution function and its graph for you to investigate.

7.7 Problems

PROBLEM 7.1. Show that the Hausdorff dimension of a countable set is equal to 0.

PROBLEM 7.2. Show that if $h > 0$, then $H(C_h) = 1$. (The set C_h is defined in Example 1.5.)

PROBLEM 7.3. Consider the Cantor function corresponding to the Cantor set $\widetilde{C}_h$.

(a) Show that the length of the graph of this Cantor function is equal to 2.
(b) Show that the Hausdorff dimension of the graph is equal to 1.
(c) Determine the length of the graph of the Cantor function corresponding to the Cantor set C_h and estimate the dimension of the graph.
(d) Show that the dimension of the graph of a nondecreasing, continuous, real-valued function, defined on $[0, 1]$, should be equal to 1.

PROBLEM 7.4. Construct a simple curve W in the unit square with $\text{area}(W) = 0$ and $H(W) = 2$.

PROBLEM 7.5. Show that the geometric comb has Hausdorff dimension equal to 1.

PROBLEM 7.6. Explain why the following statements are reasonable.

(a) The Cantor comb has dimension $1 + \ln 2/\ln 3$.
(b) If $S \subset I$ and $H(S) = c$, then $H(S \times I) = 1 + c$.
(c) The modified Cantor comb has dimension equal to $1 + \ln(2)/\ln(3)$.

PROBLEM 7.7. Show that the Hausdorff dimension of the Sierpinski triangle should be equal to $\ln(3)/\ln(2)$.

PROBLEM 7.8. Show that the Hausdorff dimension of the graph in Example 2.4 should be equal to $3/2$.

PROBLEM 7.9. Referring to Section 7.3.2, show that the simple curve through $\widetilde{C_a} \times \widetilde{C_a}$ has a tangent line at a point p on the curve $\Leftrightarrow p \notin \widetilde{C_a} \times \widetilde{C_a}$.

PROBLEM 7.10. Refer to Section 7.4.2.

(a) Determine the length of the set of points at which the Cantor function corresponding to C_h has no derivative in the calculus sense.
(b) Determine the Hausdorff dimension of the set of points at which the Cantor function corresponding to $\widetilde{C_a}$ has no derivative in the calculus sense.

PROBLEM 7.11. Can you find a simple curve that has a tangent line at each point of a set of Hausdorff dimension greater than one? Explain your answer.

Chapter 8

Julia sets and the Mandelbrot set

Gaston Julia studied the iteration of polynomials and rational functions in the early twentieth century. In complex analysis, the Julia set of a function f informally consists of those points whose long-time behavior under repeated iteration of f can change drastically under arbitrarily small perturbations. They provide striking illustrations of how a simple process can lead to intricate sets. Julia sets and the Mandelbrot set are close relatives.

8.1 Theory of Julia sets

We begin by defining a family of quadratic maps f_c from $\mathbb{C}$ to $\mathbb{C}$ by

$$f_c(z) = z^2 + c$$

for each $c \in \mathbb{C}$ and each $z \in \mathbb{C}$. As defined in Chapter 7, for $z \in \mathbb{C}$, the **orbit** of z is the sequence

$$\{z_n\} = \{f_c^{(n)}(z)\}_{n \geq 0}.$$

Definition 8.1. The **filled Julia set** J_c **for** f_c is the set of complex numbers z for which the orbit $\{f_c^{(n)}(z)\}$ is a bounded sequence. The filled Julia set is sometimes called the **prisoner** set or **resident set** for the map f_c. Its complement E_c is called the **escape set** for f_c.

Definition 8.2. The **Julia set** ∂J_c **for** f_c is the boundary of the filled Julia set for f_c; ∂J_c is also the boundary of the escape set E_c.

Thus a point in the Julia set has an orbit that does not escape to infinity, but arbitrarily nearby there are points whose orbits do escape.

Example 8.1. Let f be the map from $\mathbb{C}$ to $\mathbb{C}$ defined by $f(z) = z^2$ or $f(re^{i\theta}) = r^2e^{i2\theta}$. If $|z| < 1$, then the orbit $\{z_n\}$ of z converges to zero; and if $|z| > 1$, then

$\{z_n\}$ escapes to infinity. Finally, if $|z| = 1$, $z = e^{i\theta}$; so $|z_n| = 1$ for $n \geq 1$, and the sequence $\{z_n\}$ is a subset of the unit circle. In this case, convergence properties of $\{z_n\}$ depend on z (Problem 8.1). The filled Julia set for f is the closed unit disc $\{z : |z| \leq 1\}$ and $\partial J = \{z : |z| = 1\}$.

8.1.1 *Observations*

We make the following observations about Julia sets.

Observation 8.1. If $|z| > \max\{2, |c|\}$, then

$$|f_c(z)| - |z| \geq |z^2| - |z| - |c| = |z|\,(|z| - 1) - |c| > 0.$$

Consequently, z is in the escape set E_c for f_c; in fact (Problem 8.2), the sequence $\{|f_c{}^{(n)}(z)|\}$ increases to infinity.

Observation 8.2. Set

$$V_c = \{z : |z| > \max\{2, |c|\}\}.$$

We know that V_c is a subset of the escape set and V_c has the following additional property: If the orbit of z enters V_c, then the orbit stays in V_c, and, according to Observation 8.1, the orbit converges to infinity.

Observation 8.3. Set

$$V_{c,n} = \{z : f_c^{(n)}(z) \in V_c\}.$$

According to Observation 8.2,

$$V_c \subset V_{c,1} \subset V_{c,2} \subset V_{c,3} \subset \cdots.$$

and the escape set

$$E_c = \bigcup_{n \geq 1} V_{c,n}.$$

Observation 8.4. Set

$$F_c = \mathbb{C} \backslash V_c = \{z : |z| \leq \max\{2, |c|\}\},$$

and, for $n \geq 1$, put

$$F_{c,n} = \mathbb{C} \backslash V_{c,n} = \{z : \left|f^{(n)}(z)\right| \leq \max\{2, |c|\}\}.$$

Then

$$F_c \supset F_{c,1} \supset F_{c,2} \supset F_{c,3} \supset \cdots,$$

and each of these sets is a compact subset of $\mathbb{C}$. Moreover, if z is a root of the quadratic equation $z^2 + c = z$, then z is in each $F_{c,n}$ because $f_c^{(n)}(z) = z$ for each n. According to Proposition C.7,

$$J_c = \bigcap_{n \geq 1} F_{c,n}$$

is a nonempty compact set. The set J_c is the filled Julia set for f_c.

According to Proposition 5.6 the sequence $\{F_{c,n}\}_{n \geq 1}$ converges to J_c in the Hausdorff metric: $F_{c,n}$ is essentially indistinguishable from F_c if n is sufficiently large. The preceding remark is the reason why computer images of Julia sets and filled Julia sets represent these sets well.

8.1.2 *Visual images*

We present here some images of Julia sets for various values of the complex constant c, taken from [Weisstein (2009a)]. Pictures like those in Figure 8.1 were first seen in the late 1970s and early 1980s.

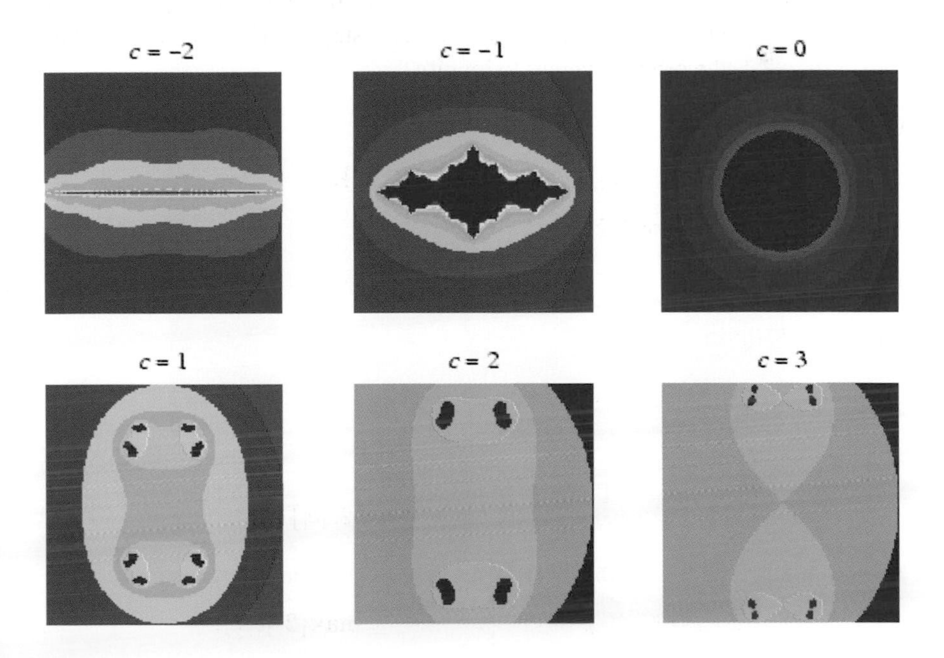

Fig. 8.1: Julia sets.

8.1.3 *Two facts about Julia sets*

The following two facts about Julia sets underlie the rest of this chapter. We state these facts without proof and recommend [Crownover (1995)] as a good source for mathematical details.

- **Fact 1:** Either J_c is connected or J_c is homeomorphic to the Cantor set which implies that the only connected subsets of J_c are one point sets.
- **Fact 2:** J_c is not connected if and only if the sequence $\{|f_c^{(n)}(0)|\}$ converges to infinity.

8.2 The Mandelbrot set

The amazing image of the Mandelbrot set shown in Figure 8.2 is taken from [Weisstein (2009b)]. Exploring its structure is a fascinating mathematical endeavor.

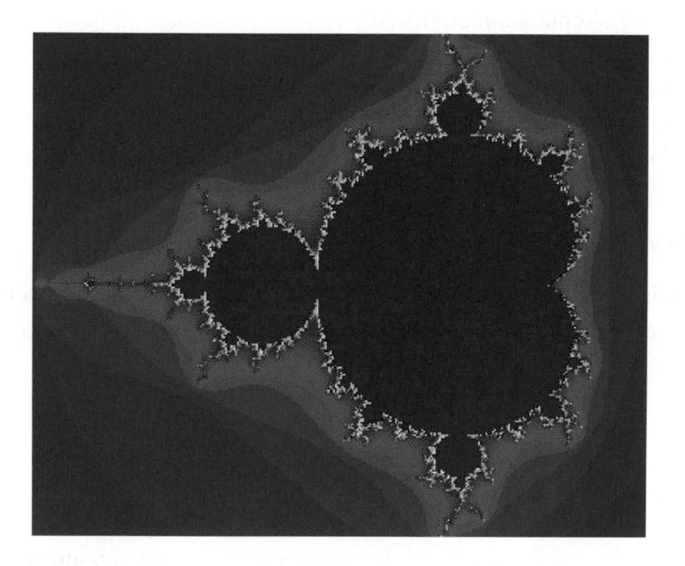

Fig. 8.2: Mandelbrot set.

Definition 8.3. The **Mandelbrot set** M is the set of complex numbers c for which J_c is a connected subset of $\mathbb{C}$.

If $|c| > 2$, then

$$|f_c(c)| = \left|c^2 + c\right| \geq \left|c^2\right| - |c| = (|c| - 1)\,|c| > |c| = \max\{2, |c|\}.$$

Consequently,

$$M = \{c \in \mathbb{C} : |c| \leq 2 \text{ and } \left|f_c^{(n)}(c)\right| \leq 2, n \geq 1\}.$$

The following project for you parallels our observations about Julia sets.

Project. For $n \geq 1$, set

$$M_n = \{c \in \mathbb{C} : \left|f_c^{(n)}(0)\right| \leq 2\}.$$

(a) Show that M_n is a closed subset of $\mathbb{C}$.
(b) Show that $M_{n+1} \subset M_n$.
(c) Show that the Mandelbrot set $M = \cap_{n \geq 1} M_n$.
(d) Show that $h(M_n, M) \to 0$ where h represents the Hausdorff metric.
(e) Describe a method for coloring $\mathbb{C}$ to display M. Explain why your method works.

The major simple geometric figure in the Mandelbrot set is the central cardioid. The second prominent simple figure is the great circle, located to the left of the central cardioid. We will use elementary properties of f_c and its iterate $f_c^{(2)}$ to explain the central cardioid and the great circle. Fixed points of f_c and its iterates play a major role in obtaining information about Julia sets and the Mandelbrot set.

8.2.1 *Fixed points of* f_c

A fixed point of f_c is in J_c. In Example 8.1, orbits of points in the interior of the unit circle converge to the fixed point 0. As defined below, 0 is an attracting fixed point and the Julia set is connected. This behavior is typical.

Definition 8.4. The point z is an **attracting fixed point** of f_c if there exists $p > 0$ such that for $w \in \mathbb{C}$ and $|w - z| \leq p$, the sequence $\{f_c^{(n)}(w)\}$ converges to z; i.e., iterates of points in $B(z, p)$ converge to (are attracted to) z.

According to the facts, $c \in M$ if f_c has an attracting fixed point because then J_c is not totally disconnected. If z is a fixed point of f_c and iterates of points in $B(z, p)$ converge to z, then $B(z, p) \subset J_c$. Our next proposition gives a sufficient condition for a point c to be in M.

Proposition 8.1. *If* $|z_c| < 1/2$*, then* $c \in M$.

Proof. Suppose z is a fixed point of $f = f_c$ and $|z| < 1/2$. Set $p = 1/2 - |z|$. We will show that the filled Julia set J_c of f_c contains the ball $B(z, p)$. Let $w \in \mathbb{C}$

with $|w-z| \leq p$. Then $|w| \leq 1/2$ and

$$\begin{aligned}|f(w)-z| &= |f(w)-f(z)| \\ &= \left|w^2-z^2\right| = |w+z|\,|w-z| \\ &\leq (|w|+|z|)\,|w-z| \leq (1-p)p.\end{aligned}$$

Since $(1-p)\,p < p$ we have that $|f(w)| \leq 1/2$.

Now, to apply mathematical induction, suppose that

$$\left|f^{(n)}(w)\right| \leq 1/2 \text{ and } \left|f^{(n)}(w)-z\right| \leq (1-p)^n p.$$

Then

$$\begin{aligned}\left|f^{(n+1)}(w)-z\right| &= \left|f^{(n+1)}(w)-f^{(n+1)}(z)\right| \\ &= \left|(f^{(n)}(w))^2-(f^{(n)}(z))^2\right| \\ &= \left|(f^{(n)}(w))^2-z^2\right| \\ &= \left|f^{(n)}(w)+z\right|\left|f^{(n)}(w)-z\right| \\ &\leq \left(\left|f^{(n)}(w)\right|+|z|\right)\left|f^{(n)}(w)-z\right| \\ &\leq (1-p)^{n+1}p.\end{aligned}$$

We conclude that the sequence $\{f^{(n)}(w)\}$ converges to z. Thus, $w \in J_c$. Consequently, $B\,(z,p) \subset J_c$. Then J_c is not totally disconnected, and $c \in M$. $\square$

The fixed points of f_c involve a square root. For any integer k the complex number $w = \sqrt{s}e^{i(\frac{\gamma}{2}+k\pi)}$ is a square root of $z = se^{i\gamma}$. The square root we use here is $\sqrt{se^{i\gamma}} = \sqrt{s}e^{i\gamma/2}$,where $-\pi < \gamma \leq \pi$. This choice for a square root function ensures that the real part of the square root of a complex number is nonnegative. We present an example before proceeding with f_c.

Example 8.2. Consider the open disk $V = \{w \in \mathbb{C} : |1-w| < 1\}$ with center and radius 1 and its boundary the circle $O = \{w \in \mathbb{C} : |1-w| = 1\}$. For $w \in O$, set $w = u + iv = re^{i\phi}$, where $-\pi/2 < \phi \leq \pi/2$, and

$$\begin{aligned}|1-w|^2 &= 1-2u+u^2+v^2 = 1, \\ u^2+v^2 &= 2u, \\ r^2 &= 2r\cos\phi, \\ r &= 2\cos\phi.\end{aligned}$$

Consequently,

$$w = re^{i\phi} = (2\cos\phi)e^{i\phi},$$

and

$$w^2 = (4\cos^2\phi)e^{i2\phi} = 2(1+\cos 2\phi)e^{i2\phi}.$$

Putting $\theta = 2\phi$, we have $-\pi < \theta \le \pi$ and

$$w^2 = 2(1+\cos\theta)e^{i\theta}.$$

The right side of the preceding equation is the locus of the boundary of a cardioid $r = 2\,(1+\cos\theta)$. (See Figure 8.3.) The map $z \to z^2$ takes the open disk V onto the interior of the cardioid and the boundary O onto the boundary of the cardioid. The equation $(1/4)\left(1-w^2\right) = 1/4-(1/2)\,(1+\cos\theta)\,e^{i\theta}$ is the scaled, reflected and translated cardioid shown in Figure 8.4.

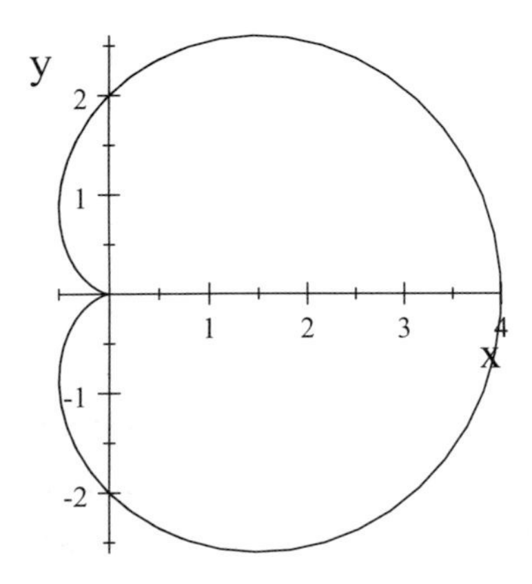

Fig. 8.3: The cardioid $2\,(1+\cos\theta)\,e^{i\theta}$.

A fixed point of f_c is a root of the quadratic equation $q(z) = z^2 - z + c = 0$. This equation has two roots

$$z = \frac{1 \pm \sqrt{1-4c}}{2}.$$

The real part of the square root term is nonnegative; thus, we choose the root with the minus sign and put

$$z_c = (1/2)(1 - \sqrt{1-4c})$$

in order to obtain a fixed point with least magnitude.

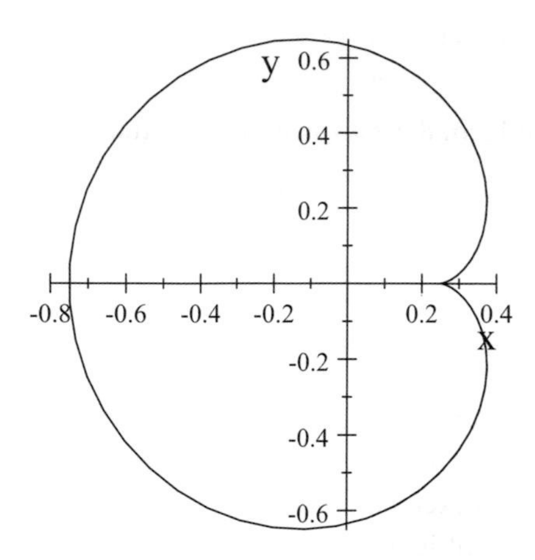

Fig. 8.4: The cardioid $1/4 - (1/2)(1 + \cos\theta)\, e^{i\theta}$.

8.2.2 *The central cardioid*

The central cardioid is the closure of the set $K = \{c \in \mathbb{C} : |z_c| < 1/2\}$. To see its shape, put $w = \sqrt{1 - 4c}$ so that $|z_c| < 1/2$ if and only if $w \in V$, and solve $w^2 = 1 - 4c$ for c to get

$$c = (1/4)(1 - w^2),$$

which is a mathematical cardioid.

8.2.3 *The great circle*

The great circle will be explained in terms of attracting period two points.

8.2.3.1 *Period two points*

Definition 8.5. A complex number z is a **period two point of** f if z is not a fixed point of f, but z is a fixed point of $f^{(2)}$; that is, $f(z) \neq z$ and $f^{(2)}(z) = z$.

Then

$$f^{(2)}(z) = f(f(z)) = (f(z))^2 + c = (z^2 + c)^2 + c.$$

So

$$\begin{aligned} f^{(2)}(z) &= (z^2 + c)^2 + c = z \Leftrightarrow \\ p(z) &= (z^2 + c)^2 + c - z = 0. \end{aligned}$$

Because fixed points of f are fixed points of $f^{(2)}$,

$$q(z) = z^2 - z + c$$

is a factor of $p(z)$. Problem 8.3 asks you to verify that

$$p(z) = h(z)q(z),$$

where

$$h(z) = z^2 + z + c + 1.$$

The condition

$$h(z) = z^2 + z + c + 1 = 0$$

is necessary for a complex number z to be a period two point of f.

The only pair c, z which solves both $q = 0$ and $h = 0$ is the pair $c = -3/4$ and $z = -1/2$. With this exception, a root z of h is a period two point.

Rewriting h in terms of its roots z_1 and z_2, we have

$$h(z) = (z - z_1)(z - z_2),$$

where

$$z_1 + z_2 = -1$$

and

$$z_1 z_2 = c + 1.$$

Applying f to a period two point z,

$$z \to f(z) \to f^{(2)}(z) = z \to f(z) \to \cdots .$$

So $f(z)$ is also a period two point. We can conclude that

$$f(z_1) = z_2, \; f(z_2) = z_1, \text{ and } z_1 \neq z_2.$$

8.2.3.2 *Description of the great circle*

Let z be a period two point of f. Let $\approx$ denote "close to" (approximate equality). For w near z, we have

$$\begin{aligned}
& f^{(2)}(w) - z \\
&= f^{(2)}(w) - f^{(2)}(z) \\
&= (f(w))^2 - (f(z))^2 \\
&= (f(w) + f(z))(f(w) - f(z)) \\
&= (f(w) + f(z))(w^2 - z^2) \\
&= (f(w) + f(z))(w + z)(w - z) \\
&\approx 2f(z)2z(w - z) \\
&= 4z_1 z_2(w - z) \\
&= 4(c + 1)(w - z).
\end{aligned}$$

These relations allow us to conclude that z_1 and z_2 comprise a pair $\{z_1, z_2\}$ of attracting period 2 points when $|4(c+1)| < 1$ or, equivalently, when

$$|c+1| < 1/4;$$

that is, when c is in the disc of radius $1/4$ with center at $(-1, 0)$. We have thus described the great circle in the Mandelbrot set M.

8.2.4 *Super-attracting fixed points*

Suppose z is a fixed point of f_c. The equalities

$$f_c(w) - z = f_c(w) - f_c(z) = w^2 - z^2 = (w+z)\,(w-z)$$

tell us that if w is "close" to z, then $f_c(w) - z = f_c(w) - f_c(z)$ is "approximately equal" to $2z(w-z)$ in the following precise sense:

$$\lim_{w \to z} \frac{f_c(w) - f_c(z)}{w - z} = \lim_{w \to z} (w+z) = 2z.$$

The limit

$$\lim_{w \to z} \frac{f_c(w) - f_c(z)}{w - z}$$

is the usual definition of the **derivative** of f at z.

As we have seen above, when $|z| < 1/2$, the magnitude $|2z|$ of the derivative $2z$ of f_c at z determines a uniform estimate for the rate of attraction of the sequence $\{f^{(n)}(w)\}$ to the attracting fixed point z. The rate of attraction is fastest when the derivative $2z = 0$.

Definition 8.6. The point z is a **super-attracting fixed point of** f_c when the derivative of f_c at z is equal to 0.

We have also seen that at a period two point z the derivative

$$\frac{d}{dz} f_c^{(2)}(z) = 4f(z)z = 4z_1 z_2 = 4(c+1).$$

Thus, when $c = -1$ is the center of the great circle, the corresponding roots $z_1 = -1$, $z_2 = 0$ of the polynomial $h(z) = z^2 + z + c + 1 = z^2 + z = 0$, are period two points of f_{-1} and super-attracting fixed points of $f_{-1}^{(2)}$.

8.2.5 *The two large bulbs adjoining the central cardioid*

We will conclude by finding the "centers" of the two large bulbs adjoining the central cardioid. These centers are values of c for which f_c has super-attracting period three points.

Period three points come in triples $\{z_1, z_2, z_3\}$:

$$z_1 \to f(z_1) = z_2 \to f(z_2) = f^{(2)}(z_1) = z_3 \to f(z_3) = z_1.$$

As in the previous section, we compute the derivative of $f^{(3)}$ at a period three point z of f:

$$\begin{aligned}
&f^{(3)}(w) - z \\
&= f^{(3)}(w) - f^{(3)}(z) \\
&= \left(f^{(2)}(w)\right)^2 - \left(f^{(2)}(z)\right)^2 \\
&= \left(f^{(2)}(w) + f^{(2)}(z)\right)\left(f^{(2)}(w) - f^{(2)}(z)\right) \\
&= \left(f^{(2)}(w) + f^{(2)}(z)\right)(f(w) + f(z))(f(w) - f(z)) \\
&= \left(f^{(2)}(w) + f^{(2)}(z)\right)(f(w) + f(z))(w + z)(w - z) \\
&\approx 2f^{(2)}(z)2f(z)2z(w - z) \\
&= 8z_3z_2z_1(w - z).
\end{aligned}$$

Thus, the triple $\{z_1, z_2, z_3\}$ is composed of super-attracting period three points when the product $8z_3z_2z_1 = 0$. We choose to label so that $z_3 = 0$; then

$$z_1 = f(z_3) = c,$$

$$z_2 = f(c) = c^2 + c,$$

and

$$\begin{aligned}
z_3 &= f(z_2) \\
&= \left(c^2 + c\right)^2 + c \\
&= c^4 + 2c^3 + c^2 + c \\
&= c(c^3 + 2c^2 + c + 1) = 0.
\end{aligned}$$

Because $c = 0$ corresponds to the super-attracting fixed point $z = 0$ and we are dealing with period three points, we seek roots of the cubic equation

$$c^3 + 2c^2 + c + 1 = 0.$$

This cubic has one real root and two complex conjugate roots. We direct your attention to Problem 8.5, which requests verification of the preceding assertions. The complex conjugate roots are the "centers" of the two large bulbs adjoining the central cardioid. The real root c_r is the "center" of a bulb to the left of the great circle.

Suppose z is one of the period three points z_1, z_2, or z_3. The derivative of $f^{(3)}(z) = f_c^{(3)}(z)$

$$\lim_{w \to z} \frac{f^{(3)}(w) - f^{(3)}(z)}{w - z} = 8f^{(2)}(z)f(z)z = 0.$$

Thus (*what follows is true, but details are omitted*), if w is close to z, then

$$\lim_{n \to \infty} f^{(3n)}(w) = f(z);$$

consequently, $c \in M$.

8.3 Generalized curves and Julia sets

We conclude with a challenging project for you. Douady and Sullivan have shown ([Sullivan (1983)], Theorem 8) that some connected Julia sets are not curves. **Project.** Can you find a connected Julia set that is not a generalized curve?

8.4 Problems

PROBLEM 8.1. Let $c = 0$ and $z = e^{it}$, so $f(z) = z^2 = e^{i2t}$. Evaluate $f^{(n)}(z)$ when $z = e^{it}$ and

(a) $t = \pi p/2^q$, where p is an odd positive integer and q is a positive integer;
(b) $t = \pi/3$;
(c) $t = \pi/5$.
(d) What can you say about the sequence $\{f^{(n)}(z)\}$ for the values of t above?
(e) Show that the sequence $\{f^{(n)}(z)\}$ is dense in the unit circle if $t = \pi u$, where u is an irrational number in $(0, 1)$. That is, show that if $w \in \mathbb{C}$ and $|w| = 1$, then a subsequence of $\{f^{(n)}(z)\}$ converges to w.

PROBLEM 8.2. Show that the sequence $\left\{\left|{f_c}^{(n)}(z)\right|\right\}$ increases to infinity if $|z| > \max\{2, |c|\}$.

PROBLEM 8.3. Verify that

$$p(z) = (z^2 + c)^2 + c - z = h(z)q(z),$$

where

$$q(z) = z^2 - z + c$$

and

$$h(z) = z^2 + z + c + 1 = 0.$$

PROBLEM 8.4. Suppose $|c + 1| < 1/4$ and z_1 and z_2 comprise a pair $\{z_1, z_2\}$ of attracting period two points for the map f_c. Show that if w is sufficiently close to z_1, then the sequence $\{f^{(2n)}(w)\}$ converges to z_1 and the sequence $\{f^{(2n-1)}(w)\}$ converges to z_2.

PROBLEM 8.5. Show that the cubic equation $c^3 + 2c^2 + c + 1 = 0$ has one real root and two complex conjugate roots. Find approximate values for these roots and locate these approximate values in the Mandelbrot set. For your approximate real root c_r of the cubic, calculate several terms in the sequence $\left\{f_{c_r}^{(n)}(0)\right\}$.

Appendix A

Points on a line

In this appendix we discuss properties of points on a line that we use in the text.

A.1 Labeling points on a line

We begin with a line L on which the integer points are labeled and proceed to label the non-zero fractional points x. Several choices of bases are used in this book so we proceed to present a labeling strategy for any base b representation of a real number where $b > 1$ is an integer. For $b = 10$, this method will label points on a line according to the familiar base 10 system. You may also be familiar with $b = 2$, the binary base 2 system.

A.1.1 base *b* *representations*

Let $x \in (0, 1)$ and $b > 1$ be an integer. For each positive integer k, let l_k denote the largest nonnegative integer for which $\frac{l_k}{b^k} \leq x$. We have

$$0 \leq l_k < b^k$$

and

$$bl_k \leq l_{k+1} < bl_k + b.$$

Put $x_1 = l_1$, $x_{k+1} = l_{k+1} - bl_k$, $y_k = \frac{l_k}{b^k}$ and $z_k = y_k + \frac{1}{b^k}$. Then

$$y_k = \frac{l_k}{b^k} = \frac{x_1}{b} + \frac{x_2}{b^2} + \cdots + \frac{x_k}{b^k} = \sum_{1 \leq j \leq k} \frac{x_j}{b^j} \leq x < y_k + \frac{1}{b^k} = z_k.$$

Define $0.x_1 x_2 \cdots x_k$ base b by

$$y_k = \frac{l_k}{b^k} = 0.x_1 x_2 \cdots x_k \text{ base } b.$$

We define a base b representation of x according to the following strategy.

If $x = \frac{l_m}{b^m}$ for some positive integer m, then there is a least integer k such that $x = 0.x_1x_2 \cdots x_k$ base b, $x_k > 0$. In this case, x has a unique, terminating base b representation. We can also write x in the computationally equivalent, (trivially) non-terminating form

$$x = 0.x_1x_2 \cdots x_k \text{ base } b = 0.x_1x_2 \cdots x_k000 \cdots \text{ base } b = 0.x_1x_2 \cdots x_k\underline{0} \text{ base } b,$$

where the *under-bar denotes continuing repetition of* 0.

Now suppose x is not of the form $\frac{l_m}{b^m}$. Then for each $k \geq 1$,

$$\frac{l_k}{b^k} = y_k < x < z_k = \frac{l_k}{b^k} + \frac{1}{b^k}.$$

In this case x has the unique, non-terminating base b representation

$$x = 0.x_1x_2 \cdots x_k \cdots .$$

Observe that if $x \in (0,1)$ has a terminating base b representation, then x is a rational number. However, not all rational numbers in $(0,1)$ have terminating base b representations as is demonstrated by the following example.

Example A.1. The number $1/2$ does not have a terminating base 3 representation. If $\frac{1}{2}$ were the left endpoint of an interval of the form $\left[\frac{l_k}{3^k}, \frac{l_k+1}{3^k}\right]$ then we would have $\frac{1}{2} = \frac{l_k}{3^k}$ for some positive integer k, suggesting the false statement that $\frac{3^k}{2}$ is an integer.

A.2 Convergence

The distance between two numbers a and b is given by

$$\text{dist}(a,b) = \text{dist}(b,a) = |b-a| = \begin{cases} b-a \text{ if } a \leq b \\ a-b \text{ if } b < a \end{cases},$$

where $|\cdot|$ denotes the absolute value function.

Example A.2. The length of an interval $[a,b] = \{x \in \mathbb{R} : a \leq x \leq b\}$ is $|b-a| = b-a$.

Definition A.1. A **sequence** in a nonempty set S is a function that assigns to each positive integer n a member y_n of S.

A sequence of real numbers can be enumerated; that is, for each positive integer n there corresponds a real number y_n. We will have need of a variety of sequences such as a sequences of functions and sequences of intervals.

Definition A.2 (Convergence of a Sequence of Points). A sequence $\{y_n\} = \{y_1, y_2, y_3, \ldots\}$ of points **converges** to a point y if (and only if) for each positive number ϵ, there exists a positive integer $K = K_\epsilon$ such that $n > K_\epsilon$ implies that $|y_n - y| \leq \epsilon$. Note that K_ϵ usually depends on ϵ.

When a sequence $\{y_m\}$ converges to y, we say that y is the *limit* of the sequence $\{y_m\}$ and write $\lim y_m = y$.

Definition A.3. A point x is **limit point** of a set S if it is the limit of a sequence of distinct points in S. The concept of a limit point extends to any metric space space setting.

A.2.1 *The geometric series*

The geometric series is a classic example of a limit. Let r denote a positive number which is less than 1. We will show that the geometric series $\sum_{i=0}^{\infty} r^i$ converges to $\dfrac{1}{1-r}$. That is, we will show that the **sequence of partial sums** $\{S_m\}_{m\geq 0}$ where

$$S_m = \sum_{i=0}^{m} r^i = 1 + r + r^2 + \cdots + r^m, m = 0, 1, \ldots.$$

satisfies $\lim S_m = \frac{1}{1-r}$. Because

$$(1-r)\, S_m = (1-r)(1 + r + r^2 + \cdots + r^m) = 1 - r^{m+1}$$

we have that

$$S_m = \frac{1 - r^{m+1}}{1-r} = \frac{1}{1-r} - \frac{r^{m+1}}{1-r}.$$

Fix $\epsilon > 0$. We need to determine $N_\epsilon \in \mathbb{N}$ so that $m > N_\epsilon$ ensures that

$$\frac{1}{1-r} - S_m = \frac{r^{m+1}}{1-r} < \epsilon \Leftrightarrow$$

$$\frac{1}{\epsilon(1-r)} < \frac{1}{r^{m+1}} = (1/r)^{m+1} = (1+s)^{m+1},$$

where $s = \frac{1-r}{r} > 0$. It suffices to choose $N_\epsilon \in \mathbb{N}$ so that

$$N_\epsilon > \frac{1}{\epsilon(1-r)s} = \frac{r}{\epsilon(1-r)^2}$$

because for $m > N_\epsilon$

$$\frac{1}{\epsilon(1-r)} < sN_\epsilon < ms \leq 1 + (m+1)\, s < (1+s)^{m+1}.$$

Thus, the sequence $\{S_m\}$ converges to $\frac{1}{1-r}$. We write

$$\sum_{i=0}^{\infty} r^i = 1 + r + r^2 + \cdots = \frac{1}{1-r}.$$

Example A.3.

$$\frac{b-1}{b^{k+1}} + \frac{b-1}{b^{k+2}} + \frac{b-1}{b^{k+3}} + \cdots = \frac{b-1}{b^{k+1}}\left(1 + \frac{1}{b} + \frac{1}{b^2} + \cdots\right) = \frac{1}{b^k}$$

since $1 + \frac{1}{b} + \frac{1}{b^2} + \cdots = \frac{b}{b-1}$.

Example A.4. Example A.3 implies that if $0 < x < 1$ has the representation

$$x = \frac{l_m}{b^m} = 0.x_1 \cdots x_{m-1}x_m \,\text{base}\, b$$

with $x_m \neq 0$, then x has the non-terminating base b representation,

$$x = 0.x_1 \cdots x_{m-1}w_m w_{m+1} \cdots = 0.x_1 \cdots x_{m-1}\,(x_m - 1)\,\underline{(b-1)}.$$

(Recall that *we use* *under-bar* *to denote continuing repetition.)*

Example A.5. The numbers $\frac{1}{4}$ and $\frac{1}{2}$ have the base 4 representations

$$\frac{1}{4} = 0.1 \,\text{base}\, 4 = 0.0\underline{3} \,\text{base}\, 4 = 0.0333 \cdots \text{base}\, 4$$

and

$$\frac{1}{2} = \frac{2}{4} = 0.2 \,\text{base}\, 4 = 0.1\underline{3} \,\text{base}\, 4$$

respectively.

Examples A.4 and A.5 illustrate the fact that a rational number of the form $\frac{i}{b^j}$ has two distinct base b representations; the first terminates and the second ends with a constant string of $(b-1)$'s.

Remark A.1. Suppose $x = 0.x_1x_2x_3 \cdots \text{base}\, b$. Then $\left(\frac{1}{b^k}\right)x = 0.\overbrace{0 \cdots 0}^{k \text{ zeros}}x_1x_2x_3 \cdots$. So multiplying a number $0 < x < 1$ by $\frac{1}{b^k}$ shifts the base b representation k places to the right.

A.3 The special nested interval property

We preface the Special Nested Interval Property with a discussion of a specific sequence of closed intervals. Let $x \in (0,1)$ have the representation $x = 0.x_1x_2x_3\cdots \text{base } b$. For each positive integer k, let l_k denote the largest nonnegative integer for which $\frac{l_k}{b^k} \le x$. Let $y_k = \frac{l_k}{b^k}$ and $z_k = y_k + \frac{1}{b^k}$. The sequence $\{[y_k, z_k]\}_{k\ge 1}$ is **nested** in the sense that

$$[y_1, z_1] \supset \cdots \supset [y_k, z_k] \supset \cdots$$

since $y_1 \le y_2 \le \cdots \le y_n \le x < z_n \le \cdots \le z_2 \le z_1$. This nested sequence of intervals has exactly one point x in its intersection. That is, x is contained in all of the intervals $[y_n, z_n]$ so that $x \in \bigcap_{n\ge 1}[y_n, z_n]$. Furthermore, if $y \in \bigcap_{n\ge 1}[y_n, z_n]$ then $|x-y| \le \lim(z_n - y_n) = 0$ which means that $x = y$ and

$$x = \bigcap_{n\ge 1}[y_n, z_n]. \tag{A.1}$$

Note that we started with x in our discussion and got the associated nested sequence $\left\{\left[\frac{l_k}{b^k}, \frac{l_k+1}{b^k}\right]\right\}_{k\ge 1}$ from x. The Special Nested Interval Property asserts that we can start with such a nested sequence and get an associated number.

Suppose $\{n_k\}_{k\ge 0}$ is a sequence of integers for which the associated sequence $\left\{\left[\frac{n_k}{b^k}, \frac{n_k+1}{b^k}\right]\right\}$ of intervals is nested:

$$[n_0, n_0+1] \supset \left[\frac{n_1}{b}, \frac{n_1+1}{b}\right] \supset \cdots \supset \left[\frac{n_k}{b^k}, \frac{n_k+1}{b^k}\right] \supset \cdots.$$

Then there is exactly one point x for which

$$\cap_{k\ge 0}\left[\frac{n_k}{b^k}, \frac{n_k+1}{b^k}\right] = x = n_0 + .x_1x_2\cdots \text{base } b;$$

n_0 is the **integer part** of x and

$$x - n_0 = 0.x_1x_2\cdots \text{base } b$$

is the **fractional part** of x. If the fractional part of x does not have a terminating base b representation then $x \in \left(\frac{n_k}{b^k}, \frac{n_k+1}{b^k}\right)$, and n_k is the largest integer for which $\frac{n_k}{b^k} \le x$.

A.4 Bounds on subsets of a line

A collection of related definitions and results about subsets of a line L follows. Several details are left for you to supply. You may simply believe the results but supplying details provides insight.

Definition A.4. Let S be a nonempty subset of a line L. A number y is a **lower bound** of S if y is less than or equal to every number in S. Similarly, z is an

upper bound of S if z is greater than or equal to every number in S. The set S is **bounded** if it is contained in some interval $[-n, n]$.

Notice that S is bounded if and only if S has a lower bound and an upper bound.

Definition A.5. A lower bound y of a nonempty set S is a **greatest lower bound** of S if y is greater than or equal to every lower bound of S. An upper bound z is a **least upper bound** of S if z is less than or equal to every upper bound of S. The least upper bound of S is denoted by lub$S$ or $\sup S$ (for **supremum**). The greatest lower bound of S is denoted by glb$S$ or $\inf S$ (for **infimum**).

The $\sup S$ is unique because if x and y are both least upper bounds for S then by definition $x \leq y$ and $y \leq x$ ensuring that $x = y$.

Theorem A.1 (The Least Upper Bound Property). *If a nonempty subset S of a line L has an upper bound, then S has a least upper bound.*

The strategy for the proof of this theorem is to define a nested sequence of intervals of the form $\left[\frac{n_k}{b^k}, \frac{n_k+1}{b^k}\right]$ each of which must contain the least upper bound of S. Then $\bigcap_{k\geq 0}\left[\frac{n_k}{b^k}, \frac{n_k+1}{b^k}\right]$ contains only one point which we will demonstrate to be $\sup S$.

Proof. Suppose S has an upper bound and define a sequence $\{n_k\}_{k\geq 0}$ by requiring that n_k be the largest positive integer for which $\frac{n_k}{b^k}$ is not an upper bound for S so that some point in S is greater than $\frac{n_k}{b^k}$. Observe that this also ensures that $\frac{n_k+1}{b^k}$ is an upper bound for S, so no point in S is greater than $\frac{n_k+1}{b^k}$. Consequently, if the least upper bound of S exists, then (for each $k \in \mathbb{N}$) the least upper bound of S must be in $\left(\frac{n_k}{b^k}, \frac{n_k+1}{b^k}\right]$.

We will now show that the intersection of the nested sequence $\left\{\left[\frac{n_k}{b^k}, \frac{n_k+1}{b^k}\right]\right\}$ of intervals is the least upper bound of S. Set

$$u = \bigcap_{k\geq 0}\left[\frac{n_k}{b^k}, \frac{n_k+1}{b^k}\right].$$

We need to verify the following two facts. First, that u is an upper bound of S and second, if $t < u$, then t is not an upper bound of S. To verify the first statement we let $w > u$ ($w \in \mathbb{R}$), and show that $w \notin S$. Since $w > u$ we know that $w \notin \bigcap_{k\geq 0}\left[\frac{n_k}{b^k}, \frac{n_k+1}{b^k}\right] = [u, u]$ so there is a positive integer k such that $w \notin \left[\frac{n_k}{b^k}, \frac{n_k+1}{b^k}\right]$. Then $w > \frac{n_k+1}{b^k}$ because $w < \frac{n_k}{b^k} \leq u$ contradicts the choice of w. Since $\frac{n_k+1}{b^k}$ is an upper bound of S, $w \notin S$. This guarantees that u is an upper bound for S.

Now let $t < u$ so that by reasoning similar to that used above $t < \frac{n_k}{b^k}$ for some positive integer k. Then t is not an upper bound of S because by construction there is an $s \in S$ such that $\frac{n_k}{b^k} < s$. So $t < s$. □

Similar statements hold for greatest lower bounds. (See Problem A.3.)

Definition A.6. A sequence $\{k_j\}$ of positive integers is **increasing** if $k_j < k_{j+1}$ for $j \geq 1$. (Note that $k_j \geq j$ for $j \geq 1$ and that we do not allow $k_j = k_{j+1}$.) If $\{y_k\}$ is a sequence and $\{k_j\}$ is an increasing sequence of positive integers, the sequence $\{y_{k_j}\} = \{y_{k_j}\}_{j\geq 1}$ is called a **subsequence** of $\{y_k\}$.

Proposition A.1. *Let $\{y_k\}$ be a sequence that converges to y. Then any subsequence of $\{y_k\}$ also converges to y.*

The proof of this proposition is left as an exercise. (See Problem A.10.)

Definition A.7. A set $E \subset S$ is **dense** in S if every point of S is a limit point of E.

A.4.1 *Bounded sequences have convergent subsequences*

Proposition A.2 shows that any bounded sequence must contain a convergent subsequence. This fact has many applications that appear throughout the book.

Definition A.8. Let S be a bounded set. If $\sup S \in S$ then we refer to $\sup S$ as the **maximum of** S and write $\max S$. In this case S contains its supremum. Similar statements can be made if S contains its infimum in which case we write $\min S$.

Proposition A.2. *Let $\{t_i\}$ be a bounded sequence. Then $\{t_i\}$ has a convergent subsequence.*

Proof. Choose a positive integer $b > 1$ to be a base. For $k \geq 0$, let n_k be the largest integer such that there are infinitely many $i \in \mathbb{N}$ for which

$$\frac{n_k}{b^k} \leq t_i.$$

Choose an increasing sequence $\{i_k\}$ of positive integers recursively such that

$$\frac{n_k}{b^k} \leq t_{i_k} < \frac{n_k + 1}{b^k}.$$

Since $\left\{\left[\frac{n_k}{b^k}, \frac{n_k+1}{b^k}\right]\right\}_{k=0}^{\infty}$ is a nested sequence of intervals with lengths converging to 0, the subsequence $\{t_{i_k}\}$ converges to the unique element $n_0+0.x_1x_2\cdots$ base b in

$$\bigcap_{k\geq 0}\left[\frac{n_k}{b^k}, \frac{n_k+1}{b^k}\right].$$

□

A.5 The real numbers $\mathbb{R}$

We started this discussion with a line L on which the integer points were labeled. We chose a positive integer $b > 1$ to be a base for a system to label the other points on the line. We represented each point on the line as a number x where $x - n_0 = 0.x_1x_2\cdots$ base b. By the real numbers, we mean the set $\mathbb{R}$ together with all the mathematical operations of addition, subtraction, multiplication, division, order, etc. Since you can translate from one value for b to another, $\mathbb{R}$ does not depend on the base used. Nor does $\mathbb{R}$ depend on the line L; however, we can identify $\mathbb{R}$ with the points on L.

A.6 Eventually periodic base *b* representations

A representation $x = 0.x_1x_2\cdots$ of x is **eventually periodic**, with period p, if there exists a positive integer m and a positive integer p such that $x_{m+i} = x_{m+i+kp}$ for any positive integers k and i. So, for example, $0.235323232\cdots$ base 6 is eventually periodic with $m = 3$ and $p = 2$. We abbreviate the periodic part by putting an under-bar under the first period and write $0.235\underline{32}$base 6.

Example A.6. We illustrate by finding the rational number $x = 0.235\underline{32}$ base 6 below. Since

$$\begin{aligned} 0.\underline{32}\text{ base }6 &= \left(\frac{3}{6}+\frac{2}{6^2}\right)\left(1+\frac{1}{6^2}+\frac{1}{6^4}+\cdots\right) \\ &= \frac{3\,(6)+2}{6^2}\left(\frac{6^2}{6^2-1}\right) \\ &= \frac{4}{7}, \end{aligned}$$

$$\begin{aligned} x &= \left(\frac{2}{6}+\frac{3}{6^2}+\frac{5}{6^3}\right)+\frac{1}{6^3}\,(0.\underline{32}\text{ base }6) \\ &= \frac{2\,(6^2)+3\,(6)+5}{6^3}+\frac{1}{6^3}\left(\frac{4}{7}\right) \\ &= \frac{223}{504} \end{aligned}$$

which is rational.

Our discussion of base b representations concludes with three simple examples converting from eventually periodic base 4 representations to fractions. The numbers in these examples will appear again in Chapter 4.

Example A.7. The value of the expansion $x = 0.\underline{21}$ base 4 is determined by the following calculations

$$\begin{aligned} 0.\underline{21}\,\text{base}\,4 &= 0.212121\cdots\text{base}\,4 \\ &= \left(\frac{2}{4}+\frac{1}{16}\right)\left(1+\frac{1}{4^2}+\frac{1}{4^4}+\cdots\right) = \frac{9}{16}\left(\frac{16}{15}\right) = \frac{3}{5}. \end{aligned}$$

Example A.8. Observe that

$$0.0\underline{21}\,\text{base}\,4 = \frac{1}{4}(0.\underline{21}\,\text{base}\,4) = \left(\frac{1}{4}\right)\left(\frac{3}{5}\right) = \frac{3}{20}.$$

Example A.9. For $x = 0.11\underline{12}$ base 4 we have

$$x = 0.111\underline{21}\,\text{base}\,4 = \frac{1}{4}+\frac{1}{16}+\frac{1}{64}+\frac{1}{64}\left(\frac{3}{5}\right) = \frac{27}{80}.$$

A.7 Problems

PROBLEM A.1. Show that a convergent sequence is bounded.

PROBLEM A.2. Verify the details of Example A.4.

PROBLEM A.3. Show that if a nonempty subset S of $\mathbb{R}$ has a lower bound, then it has a greatest lower bound.

PROBLEM A.4. Prove each of the following:

(a) Show that if $\{y_k\}$ is a bounded, non-increasing sequence, then $\{y_k\}$ converges to $\inf S$, where $S = \{y_k, k \in \mathbb{N}\}$.

(b) Let each of A and B be a nonempty subset of the real numbers $\mathbb{R}$ with the property that if $a \in A$ and $b \in B$, then $a \leq b$. Show that $\sup A \leq \inf B$.

PROBLEM A.5. For each of the following sequences $\{y_k\}$ find the associated sequence $\{[a_k, b_k]\}$. Which of these sequences converge?

(a) $y_k = \frac{(-1)^k}{k}$

(b) $y_k = \frac{1+(-1)^k}{k}$
(c) $y_k = 1 + (-1)^k + \frac{(-1)^k}{k}$
(d) $y_k = (1 + (-1)^k)\sin(\frac{k\pi}{4})$

PROBLEM A.6. Show that for any two real numbers x and y,

$$||x| - |y|| \leq |x - y|.$$

PROBLEM A.7. **[The Nested Interval Property]** Let $\{[a_k, b_k]\}_{k\geq 1}$ be a nested sequence of intervals. Show that the sequences $\{a_k\}$ and $\{b_k\}$ are bounded monotone sequences, that $\lim a_k \leq \lim b_k$, and that

$$\cap_{k\geq 1}[a_k, b_k] = [\lim a_k, \lim b_k].$$

PROBLEM A.8. **[Cauchy sequence]** A sequence $\{y_k\}$ is a **Cauchy sequence** if and only if for each $\epsilon > 0$, there exists a positive integer K_ϵ such that $m, n \geq K_\epsilon$ implies that $|y_n - y_m| < \epsilon$.

(a) Show that a Cauchy sequence is bounded.
(b) Show that a Cauchy sequence converges.

PROBLEM A.9. Prove the following:

(a) A convergent sequence is a Cauchy sequence.
(b) A sequence $\{y_k\}$ is a Cauchy sequence if for each $\epsilon > 0$, there is a positive integer p such that if $k > p$, then $|y_k - y_p| < \epsilon$.

PROBLEM A.10. Prove Proposition A.1. Specifically, verify that if $\{y_k\}$ is a sequence that converges to y, then any subsequence $\{y_{k_j}\}$ of $\{y_k\}$ also converges to y.

PROBLEM A.11. Find a convergent subsequence of each of the sequences in Problem A.5.

PROBLEM A.12. In each of the following cases, describe the set S.

(a) $S = \cap_{n\geq 1}(-1/n, 1 + 1/n)$.
(b) $S = \cap_{n\geq 1}(0, 1/n)$.
(c) $S = \cup_{n\geq 1}[1/n, 1 + 1/n]$.

PROBLEM A.13. Show that $\sqrt{2} - 1 = .0110101\cdots$ base 2.

PROBLEM A.14. Determine the four rational points in $[0, 1]$ with base 4 representations $.02\underline{21}$, $.101\underline{21}$, $.10\underline{21}$, and $.1111\underline{21}$.

PROBLEM A.15. Let $\{y_k\}$ be a bounded sequence in $\mathbb{R}$. Denote the set of subsequential limit points of $\{y_k\}$ by K:

$$K = \{y \in \mathbb{R} : \text{some subsequence } \{y_{k_j}\} \text{ of } \{y_k\} \text{ converges to } y\}.$$

(a) Show that the following three statements are equivalent.

(1) The real number z is in K.
(2) For each $\epsilon > 0$ and each positive integer N, there exists $k > N$ with $|y_k - z| < \epsilon$.
(3) For each $\epsilon > 0$, $\{k \in \mathbb{N} : |y_k - z| < \epsilon\}$ is infinite.

(b) Let $\{u_m\}$ be a sequence of points in K. Suppose that $\{u_m\}$ converges to u. Show that $u \in K$. (K is a *closed* subset of $\mathbb{R}$.)

PROBLEM A.16. Answer each of the following:

(a) Define a sequence $\{y_m\}$ of points in the interval $[0, 1]$ whose set of subsequential limit points is the whole interval $[0, 1]$.
(b) Define a sequence $\{y_m\}$ of points in $\mathbb{R}$ with the property that every real number is a subsequential limit point of the sequence $\{y_m\}$.
(c) Define a sequence $\{y_m\}$ of points in the interval $[0, 1]$ whose set of subsequential limit points is the standard Cantor set.
(d) Show that a sequence $\{y_m\}$ of positive real numbers is unbounded if and only if some subsequence of the sequence converges to infinity.

PROBLEM A.17. We construct a bounded sequence $\{y_k\}$ by defining a sequence $\{y_{n,k}\}_{k\geq 1}$ for each positive integer $n \geq 2$ and diagonalizing through the sequence of sequences that we define below.

$$\{y_{2,k}\}_{k\geq 1} = 1/2 + 1/2, 1/2 + 1/2^2, 1/2 + 1/2^3, \ldots, 1/2 + 1/2^k, \ldots$$
$$\{y_{3,k}\}_{k\geq 1} = 1/3 + 1/3, 1/3 + 1/2^2, 1/3 + 1/3^3, \ldots, 1/3 + 1/3^k, \ldots$$
$$\vdots$$
$$\{y_{n,k}\}_{k\geq 1} = 1/n + 1/n, 1/n + 1/n^2, 1/n + 1/n^3, \ldots, 1/n + 1/n^k, \ldots$$
$$\vdots$$

For $n > 1$ and $k \geq 1$,

$$y_{n,k} = 1/n + 1/n^k.$$

We obtain the sequence $\{y_k\}$ by diagonalizing through these sequences:

$y_1 = y_{2,1}$

$y_2 = y_{2,2},\ y_3 = y_{3,1},$

$y_4 = y_{2,3},\ y_5 = y_{3,2},\ y_6 = y_{4,1},$

etc. Determine the set of subsequential limit points of the sequence $\{y_k\}$, explaining why your answer is correct.

Appendix B

Length and area

We provide enough information to find lengths and areas of some unusual sets in the plane.

B.1 Intervals and length

Definition B.1. For $a < b$, an **interval** is a set E for which $(a, b) \subseteq E \subseteq [a, b]$. The length of an interval E is the non-negative real number

$$\operatorname{len}(E) = b - a.$$

Note the $\operatorname{len}([a, a]) = 0$.We extend the concept of length to additional subsets of $\mathbb{R}$.

Definition B.2. The **union** of two nonempty sets S and T is the set

$$S \cup T = \{x : x \in S \text{ or } x \in T\},$$

their **intersection** is the set

$$S \cap T = \{x : x \in S \text{ and } x \in T\},$$

and their **relative complement** is the set

$$S \backslash T = S - T = \{x : x \in S \text{ and } x \notin T\}.$$

Definition B.3. Sets S and T are **disjoint** if $S \cap T$ is the empty set.

You can verify that the intersection of two closed intervals is either empty or a closed interval. The intersection of two intervals is either empty or an interval.

Two intervals are called **nonoverlapping** if their intersection contains no open interval (i.e., is either empty or a single point set).

Definition B.4. For a finite set $\{I_j\}_{1\leq j\leq n}$ of pairwise nonoverlapping intervals

$$\operatorname{len}\left(\bigcup_{j=1}^{n} I_j\right) = \sum_{j=1}^{n} \operatorname{len}(I_j).$$

In particular, the length of a finite set of points is zero.

The following proposition indicates that definitions B.1 and B.4 are consistent.

Proposition B.1. *If an interval I is the union of n pairwise nonoverlapping subintervals $\{I_j\}_{1\leq j\leq n}$, then*

$$\operatorname{len}(I) = \sum_{j=1}^{n} \operatorname{len}(I_j). \tag{B.1}$$

Proof. We use the Principle of Mathematical Induction. Let S_n be the statement that the proposition is true for n nonoverlapping subintervals. S_1 is true. S_2 is true because we may suppose, without loss of generality, that the right end of I_1 abuts the left end of I_2 and simply add the lengths of I_1 and I_2 to get the length of I. That is, if $I = [a, b]$ so that $I_1 = [a, c]$ and $I_2 = [c, b]$ for some c, $a \leq c \leq b$, then

$$\operatorname{len}(I_1) + \operatorname{len}(I_2) = (c - a) + (b - c) = \operatorname{len}(I).$$

Now for the induction step, suppose that S_n is true, and show this supposition implies S_{n+1} is true as well. Suppose that I is the union of $n + 1$ nonoverlapping subintervals I_j, $1 \leq j \leq n + 1$. Without loss of generality, we may suppose that I_{n+1} is to the right of the other n subintervals; otherwise we may reindex the subintervals so that this is the case. The union of the first n subintervals is a subinterval J which is on the left of and abuts I_{n+1}. Apply S_n to J and the first n subintervals to get

$$\operatorname{len}(J) = \sum_{j=1}^{n} \operatorname{len}(I_j).$$

Next, apply S_2 to J and I_{n+1} to get

$$\operatorname{len}(I) = \operatorname{len}(J) + \operatorname{len}(I_{n+1}).$$

Combining these two gives

$$\operatorname{len}(I) = \sum_{j=1}^{n} \operatorname{len}(I_j) + \operatorname{len}(I_{n+1}) = \sum_{j=1}^{n+1} \operatorname{len}(I_j).$$

□

You can use Proposition B.1 to verify (See Problem B.3.) that if $\{I_j\}_{1\le j\le n}$ is a finite set of pairwise nonoverlapping intervals, each of which is a subset of an interval $[a,b]$, then

$$\operatorname{len}\left([a,b]-\bigcup_{j=1}^{n} I_j\right) = (b-a)-\sum_{j=1}^{n}\operatorname{len}(I_j). \tag{B.2}$$

B.2 Lengths of subsets of intervals

Proposition B.2, stated below without proof, describes properties of sets composed of finite unions of nonoverlapping intervals in $\mathbb{R}$.

Proposition B.2. *If each of G and H is a finite union of nonoverlapping intervals in $\mathbb{R}$, then $G\cup H$, $G\cap H$, and $G-H$ are finite unions of nonoverlapping intervals in $\mathbb{R}$.*

Definitions B.1 and B.4 are intuitive and ensure that the next two definitions are reasonable.

Definition B.5. If S is a subset of an interval $[a,b]$ and $\operatorname{len}(S)$ is defined, then

$$\operatorname{len}\left([a,b]-S\right) = \operatorname{len}\left([a,b]\right)-\operatorname{len}(S). \tag{B.3}$$

Definition B.6. Given an interval $[a,b]$ and a sequence $\{I_j\}_{1\le j<\infty}$ of pairwise nonoverlapping intervals, each of which is a subset of $[a,b]$, then

$$\operatorname{len}\left(\bigcup_{j=1}^{\infty} I_j\right) = \sum_{j=1}^{\infty}\operatorname{len}(I_j)\le b-a.$$

Example B.1 (The set of rational numbers in $[0,1]$ has length zero). An important property of the rationals $\mathbb{Q}_{[0,1]}$ in $[0,1]$ is that they can be written in the sequence

$$\{r_1,r_2,r_3,r_4,r_5,r_6,r_7,\ldots\} = \{0,1,1/2,1/3,2/3,1/4,3/4,1/5,\ldots\}.$$

The corresponding sequence of intervals

$$\{[r_1,r_1],[r_2,r_2],[r_3,r_3],\ldots\} = \{[0,0],[1,1],[1/2,1/2],\ldots\}$$

is nonoverlapping, and the length of each interval is zero. Thus, the set of rational numbers in $[0,1]$ has length equal to 0 since

$$\operatorname{len}\left(\mathbb{Q}_{[0,1]}\right) = \sum_{j=1}^{\infty}\operatorname{len}([r_j,r_j]) = 0.$$

If $\{I_j\}_{1\leq j<\infty}$ is a sequence of pairwise nonoverlapping intervals, each of which is a subset of $[a,b]$, then the equality

$$\mathrm{len}\left([a,b]-\bigcup_{j=1}^{\infty} I_j\right)=(b-a)-\sum_{j=1}^{\infty}\mathrm{len}(I_j)$$

is an immediate application of equation (B.2) coupled with Definition B.6.

Example B.2. The set of irrational numbers in $[0,1]$ has length 1.

B.3 Intervals and rectangles in the plane

The study of curves in the plane requires the concepts of intervals and rectangles in the plane.

Definition B.7. A **closed interval** I in $\mathbb{R}^2$ with endpoints (a,b) and (c,d) is the line segment connecting these two points and is often denoted by $[(a,b),(c,d)]$. The **open interval** is the line segment that does not include the endpoints and is denoted by $((a,b),(c,d))$. An **interval** I with endpoints (a,b) and (c,d) is any set that satisfies

$$((a,b),(c,d))\subset I\subset[(a,b),(c,d)].$$

Define

$$\mathrm{len}\,(I)=\sqrt{(a-c)^2+(b-d)^2}$$

which reduces to the previous definition for length whenever I lies on the real axis.

Definition B.8. Given an interval $[a,b]$ on the x-axis and an interval $[c,d]$ on the y-axis, the subset

$$[a,b]\times[c,d]$$

of $\mathbb{R}^2$ is called a **closed rectangle**. Similarly,

$$(a,b)\times(c,d)$$

is called an **open rectangle**. If

$$(a,b)\times(c,d)\subseteq E\subseteq[a,b]\times[c,d],$$

then E is called a **rectangle**.

A rectangle E has four **edges** which are subsets of intervals parallel to either the x or y axis. These edges need not have length as a subset of an interval. For example, when $E=(a,b)\times[c,d]$, $E\cap\{x=a\}$ is empty.

B.4 Length of a curve

What is the length of a curve in $\mathbb{R}^2$? To answer this question we begin with some necessary definitions.

Definition B.9. A **partition** P of an interval $[a, b]$ is a finite sequence $\{t_j\}_{0 \le j \le n}$ of points in $[a, b]$ with

$$a = t_0 < t_1 < \cdots < t_{j-1} < t_j < \cdots < t_n = b.$$

Definition B.10. The partition $Q = \{u_k\}_{0 \le k \le m}$ is a **refinement** of the partition $P = \{t_j\}_{0 \le j \le n}$ if $\{t_j : 0 \le j \le n\} \subset \{u_k : 0 \le k \le m\}$. The refinement $P \vee Q$ of P and Q is the partition whose corresponding set of points is $\{t_j : 0 \le j \le n\} \cup \{u_k : 0 \le k \le m\}$.

Given a continuous map h from $[a, b]$ to $\mathbb{R}^2$ and a partition $P = \{t_j\}_{0 \le j \le n}$, set

$$L(h, P) = \sum_{j=1}^{n} |h(t_j) - h(t_{j-1})| .$$

Remark B.2. If R is a refinement of a partition P then

$$L(h, P) \le L(h, R);$$

in particular, $L(h, P) \le L(h, P \vee Q)$ and $L(h, Q) \le L(h, P \vee Q)$. (See Problem B.1.)

For a continuous map h from $[a, b]$ to $\mathbb{R}^2$ if the set

$$\{L(h, P) : P \text{ is a partition of } [a, b]\}$$

is not bounded above, then we say that the length of $h([a, b])$ is infinite; otherwise, the length of the curve $h([a, b])$ is defined by

$$\text{len}(h, [a, b]) = \sup\{L(h, P) : P \text{ is a partition of } [a, b]\}.$$

This definition is consistent with the length of a line segment in $\mathbb{R}$ defined earlier.

B.5 Areas of subsets of the plane

We will use the sets C_h in Chapter 2 to construct sets in the plane. We need the concept of areas of subsets of $\mathbb{R}^2$ to better understand the character and structure of these sets.

B.5.1 *Areas of rectangles*

Definition B.11. The **area** of a rectangle E satisfying

$$(a, b) \times (c, d) \subseteq E \subseteq [a, b] \times [c, d]$$

is

$$\text{area}(E) = (b - a)(d - c).$$

Observe that intervals in $\mathbb{R}^2$ that are parallel to either the x or y axis have area zero even though they may have positive length. For example, area $([a, a] \times [c, d]) = 0$ but len $([a, a] \times [c, d]) = c - d$. Hence, edges of rectangles always have area equal to 0. A tedious but simple argument establishes that the area of any interval in $\mathbb{R}^2$ is zero.

Definition B.12. Two subsets of $\mathbb{R}^2$ are called **nonoverlapping** if their intersection contains no open rectangle.

Example B.3. The unit circle $T = \{(x, y) : x^2 + y^2 = 1\}$ and the unit square $U = \{(x, y) : 0 \leq x, y \leq 1\}$ are nonoverlapping. However, $D = \{(x, y) : x^2 + y^2 \leq 1\}$ and U overlap since their intersection contains, for example, the open rectangle $(0, 0.5) \times (0, 0.5)$.

If $\{E_j\}_{1 \leq j \leq n}$ is a set of pairwise nonoverlapping rectangles, then

$$\text{area}\left(\bigcup_{j=1}^{n} E_j\right) = \sum_{j=1}^{n} \text{area}(E_j).$$

As in the case of length of sets on a line, the preceding definitions of area of sets in a plane are intuitive, and analogs of Propositions B.1 and B.2 are true. The induction step to verify the analog of Proposition B.1 is more complicated than that used for length so we provide a proof.

Proposition B.3. *If a rectangle E is the union of n nonoverlapping sub-rectangles $\{E_j\}_{1 \leq j \leq n}$, then*

$$\text{area}(E) = \sum_{j=1}^{n} \text{area}(E_j).$$

Proof. Let S_n be the statement that the proposition is true for any collection of n or fewer nonoverlapping sub-rectangles. S_1 follows from the definition of the area of a rectangle. Because edges of rectangles have zero area we may assume for notational simplicity that all rectangles are closed. For the case $n = 2$ suppose

$E = [a, b] \times [c, d]$ is subdivided into $E_1 = [a, e] \times [c, d]$ and $E_2 = [e, b] \times [c, d]$. Then

$$\text{area}\,(E_1) + \text{area}\,(E_2) = ((e - a) + (b - e))\,(d - c) = \text{area}\,(E)\,.$$

Similar statements hold for the case where E is split along the y-axis. Thus, S_2 holds. Now suppose that S_n is true and that $E = [a, b] \times [c, d]$ is the union of $n + 1 \geq 3$ nonoverlapping sub-rectangles $\{E_j\}_{1 \leq j \leq n+1}$. Without loss of generality, we may suppose that the point (a, c) is on an edge of E_1 so that

$$E_1 = [a, b_1] \times [c, d_1].$$

There are 3 cases to consider. You should note that if $n + 1 = 3$ then geometric considerations verify that either Case 1 or Case 2 (or a simple modification of one of these) applies.

Case 1: $b_1 = b$**.** Here, $E_1 = [a, b] \times [c, d_1]$ so that

$$[a, b] \times [d_1, d] = \bigcup_{j=2}^{n+1} E_j.$$

We apply S_n to $\bigcup_{j=2}^{n+1} E_j$ get

$$\text{area}([a, b] \times (d_1, d]) = \sum_{j=2}^{n+1} \text{area}(E_j).$$

Now, an application of S_2 to $E_1 \cup (\cup_{2 \leq j \leq n+1} E_j)$ verifies that

$$\text{area}(E) = \text{area}\,(E_1) + \left(\sum_{j=2}^{n+1} \text{area}(E_j) \right) = \sum_{j=1}^{n} \text{area}(E_j).$$

Case 2: $d_1 = d$**.** This is similar to the first case.

Case 3: $b_1 < b$ **and** $d_1 < d$**.** This is established by first splitting E into 4 parts by intersecting it with the rectangles

$R_1 = [a, b_1] \times [c, d_1]$, $R_2 = [b_1, b] \times [c, d_1]$, $R_3 = [a, b_1] \times [d_1, d]$, and $R_4 = [b_1, b] \times [d_1, d]$,

respectively. Observe that $R_1 = E_1$. Three applications of S_2 show that

$$\begin{aligned} \text{area}(E) &= \text{area}\,((R_1 \cup R_2) \cup (R_3 \cup R_4)) \\ &= \text{area}\,(R_1 \cup R_2) + \text{area}\,(R_3 \cup R_4) \qquad \text{(B.4)} \\ &= \text{area}\,(E_1) + \sum_{k=2}^{4} \text{area}(R_k). \end{aligned}$$

Then $R_k \cap E_j, 2 \leq j \leq n+1$, do not overlap so decompose $R_k, 2 \leq k \leq 4$, into nonoverlapping rectangles according to

$$R_k = \bigcup_{j=2}^{n+1} (R_k \cap E_j) .$$

An application of S_n to this decomposition yields

$$\text{area}\,(R_k) = \sum_{j=2}^{n+1} \text{area}(R_k \cap E_j), 2 \leq k \leq 4. \tag{B.5}$$

For $2 \leq j \leq n+1$ we have that $E_j = \cup_{2 \leq k \leq 4} (R_k \cap E_j)$ since R_1 and $E_j, 2 \leq j \leq n+1$ do not overlap. Because each E_j overlaps at most two of R_2, R_3 and R_4, a modification of Case 1 or Case 2 applies to this decomposition yielding

$$\text{area}\,(E_j) = \sum_{k=2}^{4} \text{area}(R_k \cap E_j), 2 \leq j \leq n+1. \tag{B.6}$$

Summing equation B.5 over k and appealing to equation B.6 we obtain

$$\begin{aligned}
\sum_{k=2}^{4} \text{area}(R_k) &= \sum_{k=2}^{4} \sum_{j=2}^{n+1} \text{area}(R_k \cap E_j) \\
&= \sum_{j=2}^{n+1} \sum_{k=2}^{4} \text{area}(R_k \cap E_j) \\
&= \sum_{j=2}^{n+1} \text{area}\,(E_j) .
\end{aligned}$$

Using this last equation in conjunction with equation (B.4) yields the desired result.

□

Proposition B.4 extends Proposition B.2 to the plane.

Proposition B.4. *If each of G and H is a finite union of nonoverlapping rectangles in $\mathbb{R}^2$, then $G \cup H, G \cap H$, and $G - H$ are finite unions of nonoverlapping rectangles in $\mathbb{R}^2$.*

B.5.2 *Areas of general subsets of the plane*

The next two definitions are intuitive and their consistency with Definition B.11 can be established through a more rigorous analysis that is beyond the scope of this appendix. These two definitions enable us to find the areas of sets presented in the book.

Definition B.13. If S is a subset of a rectangle E and area(S) is defined, then

$$\text{area}(E - S) = \text{area}(E) - \text{area}(S). \tag{B.7}$$

Definition B.14. Suppose E is a rectangle and $\{E_j\}_{1 \leq j < \infty}$ is a sequence of pairwise nonoverlapping rectangles, each of which is a subset of E; then

$$\text{area}\left(\bigcup_{j=1}^{\infty} E_j\right) = \sum_{j=1}^{\infty} \text{area}(E_j) \leq \text{area}(E). \tag{B.8}$$

These two definitions infer that if $\{E_i\}_{1 \leq j < \infty}$ is a sequence of pairwise nonoverlapping rectangles, each of which is a subset of a rectangle E, then

$$\text{area}\left(E - \bigcup_{j=1}^{\infty} E_j\right) = \text{area}(E) - \sum_{j=1}^{\infty} \text{area}(E_j).$$

Observe that for the rectangle $E = [a, b] \times [c, d]$ we have $\text{area}\,(E) = \text{len}\,([a, b])\,\text{len}\,([c, d])$.

B.6 Problems

PROBLEM B.1. Verify the indicated inequalities in Remark B.2.

PROBLEM B.2. Given two intervals U and V in $\mathbb{R}$. Show that $\text{len}(U \cup V) = \text{len}(U) + \text{len}(V) - \text{len}(U \cap V)$.

PROBLEM B.3. Let $\{I_j\}_{1 \leq j \leq n}$ be a finite set of pairwise nonoverlapping intervals, each of which is a subset of an interval $[a, b]$. Demonstrate the validity of equation B.2.

PROBLEM B.4. Given two rectangles $U = [a, b] \times [c, d]$ and $V = [e, f] \times [g, h]$ in $\mathbb{R}^2$. Show that $\text{area}(U \cup V) = \text{area}(U) + \text{area}(V) - \text{area}(U \cap V)$.

PROBLEM B.5. Given two nonoverlapping rectangles $U = [a, b] \times [c, d]$ and $V = [e, f] \times [g, h]$ in $\mathbb{R}^2$. Show that $\text{area}(U \cap V) = 0$.

PROBLEM B.6. Given three rectangles U, V, and W in $\mathbb{R}^2$. Show that

$$\begin{aligned}\text{area}(U \cup V \cup W) &= \text{area}(U) + \text{area}(V) + \text{area}(W) \\ &\quad - \text{area}(U \cap V) - \text{area}(U \cap W) - \text{area}(V \cap W) + \text{area}(U \cap V \cap W).\end{aligned}$$

What happens if you have four rectangles?

Appendix C

Maps and sets in the plane

In this appendix we use material from Appendix A to discuss properties of maps (functions) from points on a line to points on a plane. These properties underlie much of the text.

C.1 Definition of a map

Definition C.1. A **map** (or **function**) m from D to E is a rule that assigns to each element x of D precisely one element $m(x)$ of E. Different elements of D may correspond to the same element of E. The point $m(x)$ in E is called the **image** of x. A notation used for m is

$$m : D \to E.$$

Definition C.2. The set

$$m(D) = \{m(x) : x \in D\} \subset E$$

is called the **range** of m or the **image** of D. Notice that the range of m may be a proper subset of E.

Points on a plane can be labeled by choosing a line $\mathbb{R}$ in the plane to be the x-axis and using the line which contains 0 on the x-axis and is perpendicular to the x-axis for the y-axis. A point z on the plane projects to a point x on the x-axis and a point y on the y-axis. These points are the coordinates of z. We label $z = (x, y)$, and we use the names $\mathbb{R}\times\mathbb{R}$ and $\mathbb{R}^2$ for the plane. Note that $\mathbb{R}$, the x-axis, is embedded in $\mathbb{R}^2$ since $\mathbb{R} = \{(x, 0) : x \in \mathbb{R}\}$.

Using the definition of map we see that a sequence $\{x_k\}$ of points on the x-axis is a map from **the set** $\mathbb{N}$ **of positive integers** to the x-axis so that each positive integer k is mapped onto the point x_k on the x-axis. Similarly a sequence $\{y_k\}$ of points on the y-axis is a map from $\mathbb{N}$ to the y-axis. Consequently, a sequence in $\mathbb{R}^2$ is a map from $\mathbb{N}$ into $\mathbb{R}^2$. We use the notation $\{z_k\} = \{(x_k, y_k)\}$ to indicate that k is mapped to the point $z_k = (x_k, y_k)$.

Let $D \subset \mathbb{R}$ and let m be a map from D to $\mathbb{R}$. A corresponding map G_m from D to $\mathbb{R}^2$, called the **graph map** of m, is defined by $G_m(x) = (x, m(x))$. The range of G_m is $G_m(D) = \{(x, m(x)) : x \in D\}$ and is called the **graph** of m.

C.2 Properties of points in the plane

Definition C.3. Let $z = (x, y)$ and $w = (u, v)$ be points in $\mathbb{R}^2$, and let a be a real number. Then addition and scalar multiplication are defined by

$$z + w = (x + u, y + v)$$

and

$$az = (ax, ay).$$

Subtraction is defined by

$$z - w = z + (-w).$$

Definition C.4. The **Euclidean distance** between z and w is given by

$$|z - w| = \sqrt{(x - u)^2 + (y - v)^2}. \tag{C.1}$$

In particular, the distance between $z = (x, y)$ and 0 is given by

$$|z| = |(x, y)| = \sqrt{x^2 + y^2}$$

and is called the **length** or **scale** of z.

Recall that the distance between two points x and y in $\mathbb{R}$ is $|x - y|$. A point x in $\mathbb{R}$ corresponds to the point $(x, 0)$ in $\mathbb{R}^2$, so we see that the notation $|\cdot|$ for distance in $\mathbb{R}^2$ is a consistent extension from $\mathbb{R}$ to $\mathbb{R}^2$ since

$$|s - t| = |(s, 0) - (t, 0)| = \sqrt{(s - t)^2}.$$

Geometric considerations (Figure C.1) indicate that the distance between 0 and $z + w$ is less than or equal to the sum of the distance from 0 to z and the distance from z to $z + w$:

$$|z + w| \leq |z| + |(z + w) - z| = |z| + |w|.$$

Thus,

$$|z + w| \leq |z| + |w|. \tag{C.2}$$

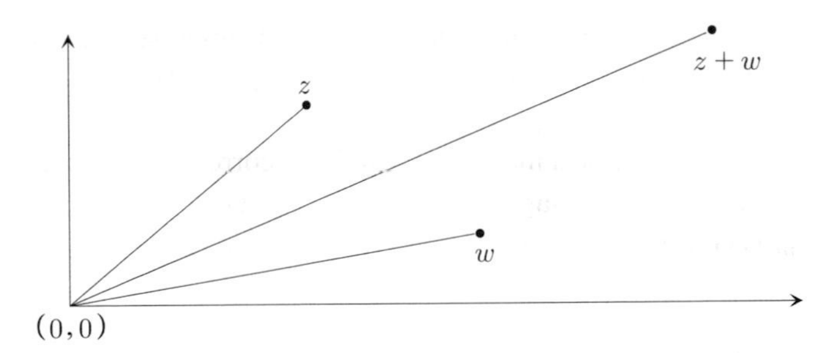

Fig. C.1: Geometric representation of the distance between points.

This inequality is called the **triangle inequality** which can be used to verify

$$|z-w| \geq ||z|-|w|| . \tag{C.3}$$

Another useful geometric property of distance in $\mathbb{R}^2$ is **homogeneity**. For any $z \in \mathbb{R}^2$ and $a \in \mathbb{R}$

$$|az| = |a|\,|z|$$

which is shown using equation (C.1). Note that homogeneity means that scalar multiplication of z by the real number a changes the length of z by $|a|$. For $z=(x,y)$ we have the inequalities

$$\max\{|x|,|y|\} \leq |z| = \sqrt{x^2+y^2} \leq |x|+|y| \leq 2\max\{|x|,|y|\} \tag{C.4}$$

that will be used frequently in this text.

Definition C.5. A sequence $\{z_n\}$ of points in $\mathbb{R}^2$ **converges** to a point z in $\mathbb{R}^2$ if for each positive number ϵ, there exists a positive integer K_ϵ such that $n > K_m$ implies that $|z_n - z| \leq \epsilon$. When $\{z_n\}$ converges to z, we say that z is the limit of the sequence $\{z_n\}$ and write

$$z_n \to z \text{ or } \lim z_n = z.$$

Proposition C.1. *A sequence $\{z_k\} = \{(x_k, y_k)\}$ of points in $\mathbb{R}^2$ is a convergent sequence in $\mathbb{R}^2$ if and only if both $\{x_k\}$ and $\{y_k\}$ are convergent sequences in $\mathbb{R}$.*

Proof. Suppose $z_k \to z = (x,y)$. Then $x_k \to x$ and $y_k \to y$ because (C.4) ensures that $|x_k - x| \leq |z_k - z|$ and $|y_k - y| \leq |z_k - z|$. Conversely, suppose $x_k \to x$ and $y_k \to y$. To show that $z_k \to z$, let $\epsilon > 0$ and choose positive integers N_x and N_y such that if $j > N_x$ then $|x_j - x| < \epsilon/2$ and $|y_k - y| < \epsilon/2$ when $k > N_y$. Applying inequalities (C.4) again we have $|z_k - z| < \epsilon$ if $k > \max\{N_x, N_y\}$. □

The definition of Cauchy sequence given in Appendix A, Problem A.8, extends naturally from $\mathbb{R}$ to $\mathbb{R}^2$. As in $\mathbb{R}$, Cauchy sequences in $\mathbb{R}^2$ converge. (See Problem C.2.)

Definition C.6. A subset S of $\mathbb{R}^2$ is **bounded** if there is a positive number B such that $|z| \leq B$ for all $z \in S$.

Proposition C.2. *A bounded sequence in $\mathbb{R}^2$ has a convergent subsequence.*

Proof. Suppose that $\{z_n\} = \{(x_n, y_n)\}$ is a bounded sequence in $\mathbb{R}^2$. According to (C.4) both $\{x_n\}$ and $\{y_n\}$ are bounded sequences in $\mathbb{R}$. By PropositionA.2 we know that a bounded sequence in $\mathbb{R}$ has a convergent subsequence. Thus, a subsequence $\{x_{n_j}\}$ of $\{x_n\}$ converges to a point x in $\mathbb{R}$. Moreover, a subsequence $\{y_{n_{j_k}}\}$ of the sequence $\{y_{n_j}\}$ converges to a point y in $\mathbb{R}$. We know that $\lim x_{n_j} = x$ implies $\lim x_{n_{j_k}} = x$. Applying Proposition C.1, the subsequence $\{z_{n_{j_k}}\}$ of $\{z_n\}$ converges to $z = (x, y)$. □

Remark C.3. According to Proposition C.2 every sequence $\{x_n\}$ in a bounded set S in $\mathbb{R}^2$ has a convergent subsequence; but if S is not bounded, then S contains sequences $\{x_n\}$ with $|x_{n+1}| > |x_n| + 1$. Such sequences have no convergent subsequences. This is an important difference between bounded and unbounded subsets of $\mathbb{R}^2$.

C.3 Continuity and limits

We are prepared to discuss pointwise continuity of maps defined on $\mathbb{R}$ or $\mathbb{R}^2$. In the following definitions D and E will denote subsets of either $\mathbb{R}$ or $\mathbb{R}^2$.

Definition C.7. The map $m : D \to E$ is **continuous** at $p \in D$ if for every positive number ϵ there is a positive number δ such that $|w - p| < \delta$, with $w \in D$, ensures that $|m(w) - m(p)| < \epsilon$. A map is **continuous on** D if it is continuous at every point of D.

Definition C.8. Suppose that p is a limit point of a set D, and a map $m : D \to E$ is given. The map m is said to have **limit** P at p if for every positive number ϵ there is a positive number δ such that $0 < |w - p| < \delta$, with $w \in D$, ensures that $|m(w) - P| < \epsilon$.

Definition C.9. Suppose that $a, t, b \in \mathbb{R}$ with $a < t < b$. Then

$$u = \lim_{s \to t^+} m(s)$$

means that the limit of $m : (t, b] \to E$ exists at t and is equal to u. Similarly

$$v = \lim_{s \to t^-} m(s)$$

means that the limit of $m : [a, t) \to E$ exists at t and is equal to v.

Theorem C.2 (A Sequential Condition for Continuity). *Let m be a map from D to E and z be a point in D. Then m is continuous at z if and only if $\{m(z_n)\}$ converges to $m(z)$ for any sequence $\{z_n\}$ in D that converges to z.*

Proof. We begin by assuming that m is continuous at $z \in D$. We must prove that $\{m(z_n)\}$ converges to $m(z)$ for any sequence $\{z_n\}$ in D that converges to z. Let $\epsilon > 0$. Because m is continuous at z, there exists $\delta > 0$ such that $w \in D$ and $|w - z| < \delta$ implies that $|m(w) - m(z)| < \epsilon$. Choose a positive integer N such that $n > N$ implies $|z_n - z| < \delta$. Thus, from the continuity of m at z we have for $n > N$, $|m(z_n) - m(z)| < \epsilon$. We conclude that $\{m(z_n)\}$ converges to $m(z)$.

Now we complete our proof by showing that if m is not continuous at z then there is a sequence$\{z_n\}$ in D that converges to z such that $\{m(z_n)\}$ does not converge to $m(z)$. Choose $\epsilon > 0$ for which there exists no $\delta > 0$ such $|m(w) - m(z)| < \epsilon$ for all $w \in D$ satisfying $|w - z| < \delta$ with $w \in D$. For each positive integer k, choose z_k with $|z_k - z| < 1/k$ and $|m(z_k) - m(z)| \geq \epsilon$. The sequence $\{z_k\}$ converges to z; but, the sequence $\{m(z_k)\}$ does not converge to $m(z)$. □

Definition C.10. A map $m : \mathbb{R} \to \mathbb{R}^2$ is represented by two maps $g, h : \mathbb{R} \to \mathbb{R}$ with $m(t) = (g(t), h(t))$.

The maps g and h in the previous definition are sometimes called the **component maps** for m. The continuity of m ensures and depends on the continuity of its component maps.

Proposition C.3. *A map $m : \mathbb{R} \to \mathbb{R}^2$ given by $m(t) = (g(t), h(t))$ is continuous at a point $t \in \mathbb{R}$ if and only if both g and h are continuous at t.*

The proof of this proposition, which relies on inequality (C.4), is Problem C.4.

C.4 Topological properties of subsets of $\mathbb{R}^2$

The domain and range of continuous maps often share the same geometric or topological properties. In such situations we say that properties are **invariant** under continuous maps. Two important invariant properties are *compactness* and *connectedness*. Compact subsets of $\mathbb{R}^2$ are those that are closed and bounded. A subset S of $\mathbb{R}^2$ is bounded if $\sup\{|p| : p \in S\} \leq M$ for some $M \in \mathbb{R}$. We continue with a discussion of general closed subsets of $\mathbb{R}^2$.

C.4.1 *Closed sets*

Definition C.11. A subset S of $\mathbb{R}^2$ is **closed** if every convergent sequence in S converges to a point in S.

It is not difficult to show that a closed interval $[a, b]$ in $\mathbb{R}$ is closed as a set according to the previous definition. So these two concepts are consistent. However, closed subsets in $\mathbb{R}^2$ include sets other than closed intervals and rectangles. For example, any finite set is closed. Also, since any convergent sequence in $\mathbb{R}$ must converge to a real number, $\mathbb{R}$ is closed. A similar statement holds for $\mathbb{R}^2$. In fact, any closed subset E of $\mathbb{R}$ is a closed subset of $\mathbb{R}^2$; E is bounded in $\mathbb{R}^2$ when E is bounded in $\mathbb{R}$. Consequently, many properties to be discussed for $\mathbb{R}^2$ apply directly to $\mathbb{R}$. The set $S = \{1/n : n \in \mathbb{N}\}$ is not closed since $0 \notin S$, but $S \cup \{0\}$ is closed. More generally, any set S in $\mathbb{R}^2$ composed of convergent sequences and their limits is closed. No bounded, non-empty segment (a, b) in $\mathbb{R}$ is closed since it is easy to construct a sequence in (a, b) that converges to a or b. (See Example C.1 below.) Notice that intervals and rectangles are bounded, while $\mathbb{R}$ and $\mathbb{R}^2$, although closed, are not bounded.

The following definition is equivalent to Definition 2.3.

Definition C.12. A set $E \subset S$ is **dense** in S if the only closed subset of S containing E is S itself.

The map f defined on the closed set $D = \{x : 1 \leq x\}$ by $f(x) = 1/x$ maps D continuously and one-to-one onto the set $(0, 1]$ which is not closed. Interestingly, the graph G_f is closed in $\mathbb{R}^2$ as verified by the following proposition.

Proposition C.4. *Let D be a closed subset of $\mathbb{R}$ and $g : D \to \mathbb{R}^2$ be continuous. Then the graph $G_g(x, g(x))$ of g is closed.*

Proof. Let g and D satisfy the stated conditions and suppose (x_n, y_n) is a convergent sequence in $G_g(x, g(x))$. Then $\{x_n\}$ is a convergent sequence in D and since D is closed its limit, say x, must also be in D. Since g is continuous on D the sequence $\{g(x_n)\}$ must converge to $g(x)$. Of course, this means that $\{G_g(x_n, g(x_n))\}$ converges to $G_g(x, g(x))$ verifying that G_g is closed. □

If S is any nonempty subset of $\mathbb{R}^2$, then the image of a constant map of S to $\mathbb{R}^2$ is a one point closed set in $\mathbb{R}^2$.

Put $S_n = \{(x,0) : x \geq n\} = [n,\infty)$. Then $\{S_n\}$ is a monotone decreasing sequence of closed sets in $\mathbb{R}^2$ with empty intersection. In contrast, according to the General Nested Interval Property presented in Problem A.7, an intersection of a sequence of nested closed intervals is a closed interval. The following proposition relates to these results.

Proposition C.5. *Let* $\{S_n\}_{n\geq 1}$ *be any sequence of closed sets in* $\mathbb{R}^2$ *with nonempty intersection. Then*

$$S = \bigcap_{n\geq 1} S_n$$

is closed.

Proof. Let $\{p_j\}_{j\geq 1}$ be any convergent sequence in S and set $p = \lim p_j$. We must show that $p \in S$. The given sequence must be in S_n for all n and since each S_n is closed, $p \in S_n$ for $n \geq 1$. Consequently, $p \in S$. □

C.4.2 *Compact sets*

Various definitions of a compact set appear in mathematical literature. Our choice for a definition of compactness permits a flexibility needed in the book.

Definition C.13. A set S is a **compact set** if every sequence of points in S has a subsequence which converges to a point in S.

Example C.1. Let $a < b$ be real numbers.

A The interval $[a,b]$ is compact. The segment (a,b) is a bounded subset of $\mathbb{R}$, but it is not closed because the sequence $\{a + \frac{b-a}{k+1}\}_{k\geq 1}$ is a sequence of points in (a,b) which converges to $a \notin (a,b)$. Consequently, (a,b) is not compact. Similarly, $(a,b]$ and $[a,b)$ are not compact.

B The set

$$\mathbb{R} - (a,b) = \{x : x \leq a\} \cup \{x : x \geq b\} = \{x \in \mathbb{R} : x \notin (a,b)\}$$

is closed, but not bounded. An unbounded set contains sequences of points which have no convergent subsequences; for example, the sequence $\{b+k\}_{1\leq k}$ has no convergent subsequence because $|b+k| \to \infty$.

As indicated by Example C.1, a compact subset of $\mathbb{R}^2$ is closed and bounded. Conversely, suppose that S is a closed and bounded subset of $\mathbb{R}^2$ and $\{x_k\}$ is a sequence of points in S. Because S is bounded, $\{x_k\}$ has a subsequence $\{x_{k_j}\}$ that converges to a point x in $\mathbb{R}^2$; x is in S because S is closed.

Proposition C.6. *A subset S of $\mathbb{R}^2$ is compact if and only if S is closed and bounded.*

The details of the proof of Proposition C.6 are left for you (see Problem C.8).

The next result strengthens the result in Proposition C.5 for the case where some S_n is compact.

Proposition C.7. *Let $E_1 \supset E_2 \supset E_3 \cdots$ be a nested sequence of nonempty closed subsets of $\mathbb{R}^2$, at least one of which, say E_m, is compact. Then*

$$E = \bigcap_{n \geq 1} E_n$$

is a nonempty compact set.

Proof. By Proposition C.5 we know that if E is not empty then it is closed. Consequently, it suffices to prove that E contains at least one point. To find a point x in the intersection E, let $x_k \in E_k$ for $k \in \mathbb{N}$. Let $\{x_{k_j}\}$ be a convergent subsequence of $\{x_k\}$ with limit x. The existence of such a subsequence and its limit are guaranteed since $\{x_{m+k}\}$ is a sequence in the compact set E_m. For each positive integer $n \geq m$, the convergent sequence $\{x_{k_j}\}_{j \geq n}$ is in the compact set E_m so x is in E_m. Consequently, x is in E establishing that E is non-empty. □

Notice that unlike Proposition C.5, Proposition C.7 does not require nontrivial intersection but rather ensures it. In addition to the assumed decreasing, nonempty nature of the sequence of sets, Proposition C.7 requires that they be closed and eventually bounded. Both of these ingredients are necessary to derive the conclusion as is illustrated in the next example.

Example C.2. We construct two collections of decreasing sets that have empty intersection.

A Put $A_n = \mathbb{R} - (-n, n)$. Each A_n is a closed subset of $\mathbb{R}$, and $\{A_n\}$ is a decreasing sequence of sets. However, $\cap_{n \geq 1} A_n$ is empty (i.e., there is no point which is in all of the sets) because for any real number x there is a positive integer n such that $-n \leq x \leq n$ so that $x \notin A_n$. Note that each A_n has infinite length and is unbounded.

B Let A_n denote the segment $(0, 1/n)$. Then $\{A_n\}$ is a decreasing sequence of bounded subsets of $\mathbb{R}$. Also, $\cap_{n\geq 1} A_n$ is empty because for any $x > 0$ there is a positive integer n such that $\frac{1}{n} < x$ which means that $x \notin A_n$. Observe that each A_n is bounded but not closed.

Theorem C.3. *Let m be a continuous map from a compact set $S \subset \mathbb{R}^2$ to $\mathbb{R}^2$. Then the image $m(S)$ of S is compact.*

Proof. Let m be a continuous map from a compact set S to $\mathbb{R}^2$ and suppose that $\{y_n\}$ is a sequence of points in $m(S)$. For each positive integer n let $x_n \in S$ be so that $m(x_n) = y_n$. Then the sequence $\{x_n\}$ in S has a subsequence $\{x_{n_k}\}$ which converges to, say, $x \in S$ since S is compact. Because m is continuous the sequence $\{y_{n_k} = m(x_{n_k})\}$ converges to $m(x) \in m(S)$. Thus, $m(S)$ is compact since the original sequence $\{y_n\}$ in $m(S)$ has a convergent subsequence which converges to a point in $m(S)$. □

Definition C.14. The **distance between two disjoint closed sets** S and T in $\mathbb{R}$ or $\mathbb{R}^2$ is defined by the formula

$$d(S,T) = \inf\{|s-t| : s \in S, t \in T\}.$$

It is possible for two non-intersecting, closed sets S and T to have zero distance between them as the next example demonstrates.

Example C.3. Let $S = \{n : n \in \mathbb{N}\}$ and $T = \left\{n + \frac{1}{n+1} : n \in \mathbb{N}\right\}$. Both S and T are closed because neither contains a convergent sequence except for those with a constant tail in which case the sequence converges to this constant. Also, $S \cap T$ is empty so they are disjoint. But

$$0 \leq d(S,T) \leq \inf\left\{\left|n - \left(n + \frac{1}{n+1}\right)\right| : n \in \mathbb{N}\right\} = 0.$$

Notice that neither of the sets in Example C.3 is compact because they are unbounded. The next proposition provides additional information concerning the distance between two closed sets when at least one of the sets is compact.

Proposition C.8. *If S and T are nonempty, disjoint, closed subsets of $\mathbb{R}^2$ with S compact, then $d(S,T) > 0$. Moreover, there is a point $s \in S$ such that*

$$d(S,T) = d(\{s\}, T) = d(s,T). \tag{C.5}$$

Proof. Suppose S and T are nonempty subsets of $\mathbb{R}^2$ where S is compact and T is closed. We will establish the result by showing that $d(S,T) = 0$ implies that S and T are not disjoint. If $d(S,T) = 0$ there is a sequence $\{\{s_k, t_k\} \in S \times T\}$

for which $|s_k - t_k| < 1/k$, $k \in \mathbb{N}$. Since S is compact, a subsequence $\{s_{k_j}\}$ of the sequence $\{s_k\}$ converges to a point $s \in S$; the corresponding subsequence $\{t_{k_j}\}$ also converges to s because $|s_k - t_k| \to 0$. Since T is closed $s \in T$. Thus, $s \in S \cap T$ so these sets are not disjoint.

To prove the second part of the proposition we assume that $d(S,T) > 0$ and modify the construction of s given in the first part. Specifically, let $\{s_k\}$ be a sequence of points in S that satisfy

$$|d(s_k, T) - d(S,T)| < \frac{1}{k}$$

for each positive integer k. Then, as before, $\{s_k\}$ must contain a subsequence that converges to a point $s \in S$ because S is compact. The point s satisfies equation (C.5). □

Definition C.15. The **cross product** of a pair of sets S and T is the set

$$S \times T = \{(z,w) : z \in S \text{ and } w \in T\}.$$

Proposition C.9. *The cross product of two compact subsets of* $\mathbb{R}^2$ *is compact.*

Proof. Let S and T be compact subsets of $\mathbb{R}^2$ and let $\{(z_n, w_n)\}$ be a sequence of points in $S \times T$. Then $\{z_n\}$ is a sequence in the compact set S and thus has a convergent subsequence $\{z_{n_j}\}$ which converges to $z \in S$. Likewise, the sequence $\{w_{n_j}\}$ in T has a convergent subsequence $\left\{w_{n_{j_k}}\right\}$ which converges to a point w in T. Hence, the subsequence $\left\{\left(z_{n_{j_k}}, w_{n_{j_k}}\right)\right\}$ of $\{(z_n, w_n)\}$ must converge to $(x,w) \in S \times T$. □

(The distance between two points (x,w) and (u,v) in $\mathbb{R}^2 \times \mathbb{R}^2$ is $\sqrt{|x-u|^2 + |w-v|^2}$.)

Corollary C.1. *If* S *and* T *are disjoint compact subsets of* $\mathbb{R}^2$ *then there are points* $s \in S$ *and* $t \in T$ *such that* $d(S,T) = |s-t|$.

Proof. The map $d : S \times T \longrightarrow \mathbb{R}$ is continuous. Because the continuous image of a compact set is compact, d attains its minimum on $S \times T$. □

C.4.3 *Connected sets*

We develop the definition of connectedness of subsets of $\mathbb{R}^2$ by first defining the concept of disconnected sets. Various equivalent definitions of a general disconnected set can be given. A definition valid for closed sets is given below; it suffices for our purposes because we only need to discuss connectedness of closed sets. A general definition of connected appears in Problem C.17.

Definition C.16. A closed set in $\mathbb{R}^2$ **is disconnected** if it can be separated into two nonempty, disjoint, closed subsets.

Definition C.17. A set S in $\mathbb{R}^2$ is **connected** if it is not disconnected. The largest connected subset of S containing a point x is called a **connected component of** x. The set S is **totally disconnected** if the connected component of each point consists of just that point. (See [Falconer (1990)].)

Example C.4. According to this definition the set $[0,1] \cup [2,3]$ is clearly disconnected since it can be split into the two non-empty, disjoint, closed subsets $[0,1]$ and $[2,3]$.

Example C.5. Set $S = \{(x, 1/x) : x \neq 0\}$, $T = \{(x,y) : xy = 0\}$ and $U = S \cup T$. The set U is disconnected and S and T are disjoint, closed sets. The set S is disconnected since it is composed of two branches of a hyperbola. Finally, T is connected.

Example C.6. The set of positive integers is totally disconnected.

Any set containing a single point in $\mathbb{R}^2$ is connected. However, a finite set containing more than one point is disconnected.

Proposition C.10. *A compact subset J of $\mathbb{R}$ is connected if and only if J is an interval.*

Proof. Let J be a compact subset of $\mathbb{R}$. Put $u = \inf(J)$ and $v = \sup(J)$. Such numbers are available to us because J is bounded. Because J is closed, $u, v \in J$. Notice that $J \subset [u,v]$. Two steps remain.
Step 1. We show that if $J \neq [u,v]$, then J is disconnected. Suppose that $t \in (u,v)$ and $t \notin J$. Then $[u,t] \cap J$ and $[t,v] \cap J$ are non-empty, disjoint closed sets whose union is J.
Step 2. We show that if J is not connected, then J is not an interval. Suppose J is not connected. Then there exist non-empty, disjoint, closed sets A and B whose union is J. There exist points $a \in A$ and $b \in B$ such that $|a - b| = d(A,B) > 0$. Consequently, the midpoint of a and b, $(a+b)/2 \notin J$, so J is not an interval. □

A slight modification of preceding proof shows (Problem C.10) that for $r > 0$, $\{z \in \mathbb{R}^2 : |z| \leq r\}$ is a connected set.

Proposition C.11. *Suppose A and B are disjoint compact sets with $d(A,B) = 2p > 0$. Suppose that E is a compact, connected set that contains $A \cup B$. Then there exists $z \in E$ such that $d(z, A \cup B) \geq p$.*

Proof. For $0 \leq \delta < p$, put $A_\delta = \{w : d(w, A) \leq \delta\}$ and $B_\delta = \{w : d(w, B) \leq \delta\}$. Then A_δ and B_δ are disjoint compact sets. Because E is connected and contains $A \cup B$, E is not contained in $A_\delta \cup B_\delta$. Thus, for $\delta_n = \left(\frac{n}{n+1}\right) p$, we choose $z_n \in E \backslash (A_{\delta_n} \cup B_{\delta_n})$. Let $z_n \to z$. Then $d(z, A \cup B) \geq p$. □

Proposition C.12. *Let $\{E_k\}$ be a nested sequence of compact, connected sets. Then $E = \cap_{k \geq 1} E_k$ is a compact, connected set.*

Proof. Let A and B be disjoint compact subsets of E. For $n \geq 1$, choose $z_n \in E_n$ such that $d(z_n, A \cup B) \geq (1/2)\, d(A, B) = p > 0$. Then there is a subsequence $\{z_{n_k}\}$ such that $z_{n_k} \to z$. Then $z \in E$ and $d(z, A \cup B) \geq p$, so $E \neq A \cup B$. □

Proposition C.13. *The continuous image of a compact, connected set in $\mathbb{R}^2$ is a compact, connected set.*

Proof. Let D be a compact subset of $\mathbb{R}^2$. Let m be a continuous map from D to $\mathbb{R}^2$ and $E = m(D)$ denote the image of D. We know that E is compact because it is the continuous image of a compact set (Theorem C.3). We will show that if E is not connected, then D is not connected. Suppose that $E = F \cup G$, where F and G are nonempty, disjoint compact sets. Let $A = \{x \in D : m(x) \in F\}$ and let $B = \{x \in D : m(x) \in G\}$. Suppose we are given a convergent sequence in A. This sequence must converge to some point, say x, in D because D is closed. But then $m(x)$ must be in F since $d(F, G) > 0$ so $x \in A$. Thus, A is closed. Likewise, B is closed. Consequently, D is not connected because A and B are disjoint, compact sets whose union is D. The result follows. □

Proposition C.14. *Let m be a continuous map from a closed interval $[a, b]$ to $\mathbb{R}$. Then the image of m is a closed interval.*

Proposition C.14 is an immediate consequence of Propositions C.10 and C.13.

Curves, the continuous images of I in $\mathbb{R}^2$, are a central focus of this book. According to Proposition C.13 curves are compact and connected subsets of $\mathbb{R}^2$.

C.4.4 *Fixed points of maps*

Fixed point theorems have an instrumental role in studying fractal images and have numerous applications in theoretical and applied mathematics. We will apply Proposition C.14 to verify a fixed point theorem (see Theorem C.4).

Definition C.18. Let m be a map from a set S to S. A point x in S is called a **fixed point** of m if $m(x) = x$.

Theorem C.4. *Let m be a continuous map from an interval $[a, b]$ to $[a, b]$. Then m has a fixed point.*

Proof. Put $f(x) = x - m(x)$. Then $f(a) \leq 0$ and $f(b) \geq 0$. Thus, there exists $x \in [a, b]$ with $f(x) = 0$. □

C.4.5 *Uniform continuity of maps*

Suppose m is a *continuous* map from $S \subset \mathbb{R}^2$ to $\mathbb{R}^2$. Then for any w in S and $\epsilon > 0$ there is $\delta > 0$ such that $z \in S$ and $|z - w| < \delta$ implies that $|m(z) - m(w)| < \epsilon$. *Here δ normally depends on both w and ϵ.* Uniform continuity is a more restrictive concept that permits you to eliminate the dependence of δ on which point in the domain is being investigated.

Definition C.19. The map m is **uniformly continuous** on S if for each $\epsilon > 0$, you can choose a δ such that $z, w \in S$ with $|z - w| < \delta$ ensures that $|m(z) - m(w)| < \varepsilon$. Notice that δ depends only on ϵ and works throughout S. (The last two inequalities can be changed to $\leq$.)

Proposition C.15. *A continuous map from a compact subset S of $\mathbb{R}^2$ to $\mathbb{R}^2$ is uniformly continuous on S.*

Proof. Let S be a compact subset of $\mathbb{R}^2$ and let m be a continuous map from S to $\mathbb{R}^2$. We will show that if m is not uniformly continuous on S, then m is not continuous on S. Suppose that m is not uniformly continuous on S. Then there is an $\epsilon > 0$ and a sequence$\{\{z_k, w_k\}\}_{k \geq 1}$ of pairs of points in S with $|z_k - w_k| < \frac{1}{k}$ and $|m(z_k) - m(w_k)| \geq \epsilon$. Let $\{\{z_{k_j}, w_{k_j}\}\}_{j \geq 1}$ be a subsequence for which the sequence $\{z_{k_j}\}$ converges to a point z in S. Since $|z_k - w_k| < \frac{1}{k}$, the sequence $\{w_{k_j}\}$ also converges to z. However, $\left|m(z_{k_j}) - m\left(w_{k_j}\right)\right| \geq \epsilon$. Thus, m is not continuous at w. (Problem C.13 asks you to provide details of this last remark.) □

There are uniformly continuous maps on non-compact sets. The map $f(z) = z$ is uniformly continuous on the set $\{z = (x, y) : x, y > 0\}$ which is neither bounded nor closed. As shown by the next example compactness is necessary for ensuring the uniform continuity of a continuous map.

Example C.7. Observe that in both parts below m is continuous on its domain.

A Let $m(x) = 1/x$ for $x \in (0,1]$. Since $|1/(n+1) - 1/n| = 1/(n^2+n) < 1/n \longrightarrow 0$ and $m(1/(n+1)) - m(1/n) = 1$ we see that for $\epsilon > 0$ there is no fixed δ that satisfies

$$|x - y| < \delta \Longrightarrow |m(x) - m(y)| < \epsilon$$

for $x, y \in (0,1]$. Thus, m is not uniformly continuous on the bounded but not closed set $(0,1]$.

B Let $m(x) = x^2$ for $x \in [1,\infty)$. Then, $|(n+1/n) - n| = 1/n \to 0$ and $m(n+1/n) - m(n) = 2 + 1/n^2 > 2$. Similar reasoning shows that m is not uniformly continuous on the closed but unbounded set $[1,\infty)$.

C.5 Convergence of maps

Some curves in $\mathbb{R}^2$ can be effectively represented as the image of the limit of a sequence of continuous maps from I to $\mathbb{R}^2$. To define such a limit we need the concept of *uniform convergence* for a sequence of maps. A sequence of maps $\{f_k\}$ defined on a closed interval may exhibit types or "strengths" of convergence.

Definition C.20 (Convergence for sequences of continuous maps). Let $\{f_k\}$ be a sequence of continuous maps from an interval $[a,b]$ to $\mathbb{R}^2$.

(1) The sequence $\{f_k\}$ **converges at a point** x in $[a,b]$ if $\lim f_k(x)$ exists.

(2) The sequence $\{f_k\}$ **converges pointwise on** $[a,b]$ if $\lim f_k(x)$ exists at each point x in $[a,b]$. The sequence $\{f_k\}$ **converges pointwise to** f **on** $[a,b]$ if $\lim f_k(x) = f(x)$ at each point x in $[a,b]$.

(3) The sequence $\{f_k\}$ **converges uniformly to** f **on** $[a,b]$ if for each $\epsilon > 0$ there exists a positive integer K_ϵ such that $k \geq K_\epsilon$ and $x \in [a,b]$, imply that $|f_k(x) - f(x)| \leq \epsilon$. *Note that K_ϵ depends on ϵ but not on x.*

Observe that these three concepts of convergence increase in strength as we progress through the definitions in the sense that for a given sequence of maps defined on an interval, uniform convergence implies pointwise convergence on the interval which, in turn, implies convergence at a point in the interval.

Example C.8. The following illustrate these types of convergence.

A. Convergence at a point Let $f_k(x) = kx$ for $x \in [0,1]$. Then clearly the sequence of real numbers $\{f_k(0)\}$ converges to 0. However, for fixed $x \in$

$(0, 1]$ and $N \in \mathbb{N}$, $f_n(x) > N$ if $n > N/x$. (See Proposition A.1.) Thus $\{f_n(x)\}_{n \geq 1}$ converges only when $x = 0$.

B. Point-wise convergence In this construction we define $f_k(x)$ on $[0, 1]$ for each positive integer k by

$$f_k(x) = \begin{cases} kx & \text{if } 0 \leq x \leq 1/k \\ -kx + 2 & \text{if } 1/k < x < 2/k \\ 0 & \text{if } 2/k \leq x \leq 1 \end{cases}. \tag{C.6}$$

The graphs of f_4, f_5, and f_6 are depicted in Figure C.2. First, we claim that $\{f_k(x)\}$ converges to $f(x) = 0$ for any $x \in [0, 1]$. Clearly, $\{f_k(0)\}$ converges to 0 so assume $x \in (0, 1]$. The desired convergence for x follows immediately because $f_k(x) = 0$ whenever $k \in \mathbb{N}$ satisfies $\frac{2}{k} < x$. Notice however that this sequence does not converge uniformly to $f(x) = 0$ since

$$\sup_{x \in (0,1]} \{|f_k(x) - f(x)|\} = f_k\left(\frac{1}{k}\right) = 1.$$

You may wish to construct a simple modification of these maps for which

$$\sup_{x \in (0,1]} \{|f_k(x) - f(x)|\} = \infty.$$

C. Uniform convergence If we set $g_k(x) = \frac{1}{k} f_k(x)$ for each positive integer k where f_k is given by equation C.6 then

$$|g_k(x) - g(x)| < \frac{1}{k}$$

for all $x \in [0, 1]$ where $g(x) = 0$. So g_k converges to g uniformly on I.

As mentioned previously, uniform convergence is the most demanding of the three presented in Definition C.20. The first two types of convergence depend on the point chosen. Uniform convergence is independent of the points in the interval and can therefore be thought of as convergence of the maps themselves. Consequently, it is appropriate to write $f_k \longrightarrow f$ when the sequence $\{f_k\}$ converges uniformly to f.

For a bounded map g on I, put

$$\|g\| = \|g\|_{[0,1]} = \sup_{x \in [0,1]} |g(x)|.$$

Then

$$\|f_k - f\|_{[a,b]} = \sup_{x \in [a,b]} (|f_k(x) - f(x)|).$$

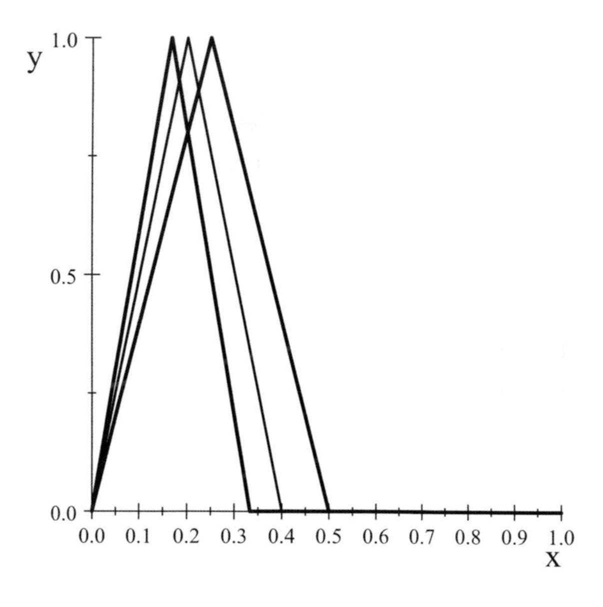

Fig. C.2: Graphs of $f_k(x)$ given in Example C.8B.

An equivalent statement of uniform convergence of $\{f_k\}$ to f on $[a,b]$ is

$$\lim_k \|f_k - f\|_{[a,b]} = 0.$$

This notation emphasizes that uniform convergence, although depending on an interval, does not depend on any point in the interval. When the context justifies a reduction in notation we simply write $\|f_k - f\|$ for $\|f_k - f\|_{[a,b]}$.

Theorem C.5. *Let $\{f_k\}$ be a sequence of continuous maps from $[a,b]$ to $\mathbb{R}^2$ which converges uniformly to f on $[a,b]$. Then f is continuous on $[a,b]$.*

Proof. Fix $x \in [a,b]$. We will show that $f([a,b])$ is continuous at x. Let $\epsilon > 0$. Since $\lim \|f_k - f\| = 0$, we can choose a positive integer k such that

$$\|f_k - f\| < \epsilon/3.$$

After choosing k and relying on the uniform continuity of f_k we can choose a $\delta > 0$ such that if $y \in [a,b]$ and $|y - x| < \delta$ then

$$|f_k(y) - f_k(x)| < \epsilon/3.$$

For such y we have using the triangle inequality

$$\begin{aligned}
&|f(y) - f(x)| \\
&= |(f(y) - f_k(y)) + (f_k(y) - f_k(x)) + (f_k(x) - f(x))| \\
&\leq |f(y) - f_k(y)| + |f_k(y) - f_k(x)| + |f_k(x) - f(x)| \\
&< \frac{\epsilon}{3} + \frac{\epsilon}{3} + \frac{\epsilon}{3} = \epsilon.
\end{aligned}$$

Thus, f is continuous at x. Since x is an arbitrary point in $[a, b]$, f is continuous on $[a, b]$. □

Definition C.21. The statement that the sequence of maps $\{f_k\}$ is **uniformly Cauchy on** $[a, b]$ means that for each $\epsilon > 0$ there exists a positive integer K_ϵ such that if $j \geq k \geq K_\epsilon$ then $\|f_j - f_k\|_{[a,b]} \leq \epsilon$.

Theorem C.6. *Let $\{f_k\}$ be a uniformly Cauchy sequence of continuous maps from $[a, b]$ to $\mathbb{R}^2$. There is a map f on $[a, b]$ such that the sequence $\{f_k\}$ converges uniformly to f on $[a, b]$.*

Proof. Since the sequence is uniformly Cauchy it is pointwise Cauchy. So for each x in $[a, b]$, the sequence $\{f_k(x)\}$ converges to a number which we denote by $f(x)$. This correspondence between a number x and $f(x)$ defines a map f on $[a, b]$ since the limit of a convergent sequence of numbers is unique. To show that $\|f_k - f\| \longrightarrow 0$ as k increases without bound, let $\epsilon > 0$ and choose a positive integer K_ϵ such that if m and n are greater than or equal to K_ϵ, then $\|f_m - f_n\| \leq \epsilon$. Let $x \in [a, b]$. Then, for $m \geq K_\epsilon$,

$$\begin{aligned}
|f_m(x) - f(x)| &= \lim_{n \to \infty} |f_m(x) - f_n(x)| \leq \sup_{n \geq k} |f_m(x) - f_n(x)| \\
&\leq \sup_{n \geq K_\epsilon} \|f_m - f_n\| \leq \epsilon.
\end{aligned}$$

Since x is an arbitrary point in $[a, b]$, we conclude that

$$\|f_m - f\| \leq \epsilon$$

whenever m is a positive integer $\geq K_\epsilon$. Thus, $\{f_k\}$ converges uniformly to f on $[a, b]$. □

Corollary C.2. *By Theorem C.5 the limit of a uniformly Cauchy sequence of continuous maps is continuous.*

Theorem C.7. *Let $\{f_k\}$ be a sequence of continuous maps from $[a, b]$ to $\mathbb{R}^2$. Suppose there exist positive numbers u and U with $U < 1$ such that $\|f_{k+1} - f_k\| \leq uU^k$ for all positive integers k. Then the sequence $\{f_k\}$ is uniformly Cauchy on $[a, b]$.*

Proof. The inequalities

$$\begin{aligned}
&\|f_{k+m} - f_k\| \\
&\le \|f_{k+m} - f_{k+m-1}\| + \|f_{k+m-1} - f_{k+m-2}\| + \cdots + \|f_{k+1} - f_k\| \\
&\le uU^{k+m-1} + uU^{k+m-2} + \cdots + uU^{k+1} + uU^k \\
&= uU^k(1 + U + \cdots + U^{m-1}) \\
&< \left(\frac{u}{1-U}\right) U^k
\end{aligned}$$

verify the desired nature of $\{f_k\}$ since $\lim_{k\to\infty} \left(\frac{u}{1-U}\right) U^k = 0$. □

Theorems C.5, C.6, and C.7 describe situations where a sequence $\{f_k\}$ of continuous maps from $[a, b]$ to $\mathbb{R}^2$ converges uniformly to a continuous map f on E. As can be seen from their proofs, these results easily extend to the case where the domain of the maps in the sequence is a non-empty subset E of $\mathbb{R}^2$. For example, we use these results when the domain of the maps is a Cantor set.

Theorem C.8. *Let $\{f_n\}$ be a sequence of continuous maps from a non-empty set $E \subseteq \mathbb{R}^2$ to $\mathbb{R}^2$. Suppose the sequence $\{f_n\}$ converges uniformly to f on E. Let $x \in E$. Suppose $\{x_n\}$ is a sequence of points in E and $x_n \to x$. Then*

$$f_n(x_n) \to f(x).$$

Proof. Theorem C.5 tells us that f is continuous on E. Let $\epsilon > 0$. Choose $\delta > 0$ such that if $y \in E$ and $|y - x| < \delta$ then $|f(y) - f(x)| < \epsilon/2$. Choose $N_f \in \mathbb{N}$ such that if $n > N_f$ then $\|f_n - f\| < \epsilon/2$. Choose $N_{\text{seq}} \in \mathbb{N}$ such that if $n > N_{\text{seq}}$ then $|x_n - x| < \delta$. Put $N = \max\{N_f, N_{\text{seq}}\}$. For $n > N$,

$$\begin{aligned}
|f_n(x_n) - f(x)| &= |f_n(x_n) - f(x_n) + f(x_n) - f(x)| \\
&\le |f_n(x_n) - f(x_n)| + |f(x_n) - f(x)| \\
&\le \|f_n - f\| + |f(x_n) - f(x)| \\
&< \epsilon/2 + \epsilon/2 = \epsilon.
\end{aligned}$$

□

C.6 Linear maps from $\mathbb{R}^2$ to $\mathbb{R}^2$

A map $M : \mathbb{R}^2 \to \mathbb{R}^2$ is **linear** if

$$M(az + w) = aM(z) + M(w)$$

for all $a \in \mathbb{R}$, and $z, w \in \mathbb{R}^2$. Let $A = A(\mathbb{R}^2)$ denote the set of linear maps (operators) $M : \mathbb{R}^2 \to \mathbb{R}^2$. Each map M in A has a representation as a 2×2

matrix. To see this let M be a linear operator and suppose that $M((1,0)) = (a,b)$ and $M((0,1)) = (c,d)$. Then for $z = (x,y) \in \mathbb{R}^2$ we have

$$\begin{aligned} M((x,y)) &= xM((1,0)) + yM((0,1)) \\ &= (ax+cy, bx+dy). \end{aligned}$$

Using matrix notation we have

$$\begin{bmatrix} a & c \\ b & d \end{bmatrix} \begin{bmatrix} x \\ y \end{bmatrix} = \begin{bmatrix} ax+cy \\ bx+dy \end{bmatrix}.$$

Consequently, M is represented by the matrix $\begin{bmatrix} a & c \\ b & d \end{bmatrix}$ and we write

$$M \cong \begin{bmatrix} a & c \\ b & d \end{bmatrix}.$$

If N is a linear operator with matrix representation

$$N \cong \begin{bmatrix} \alpha & \delta \\ \beta & \gamma \end{bmatrix},$$

then the **matrix product** MN is defined by

$$MN \cong \begin{bmatrix} a\alpha + c\beta & a\delta + c\gamma \\ b\alpha + d\beta & b\delta + d\gamma \end{bmatrix}.$$

Linear operators M and N satisfy

$$\begin{aligned} (M+N)(w) &= M(w) + N(w), \text{ and} \\ (MN)(w) &= M(N(w)). \end{aligned}$$

We can use matrix maps to achieve some desired geometric operations.

(1) Counter-clockwise rotation by an angle θ in the plane is given by $z \to M_\theta z$ where the matrix representation of M_θ is

$$M_\theta = \begin{bmatrix} \cos\theta & -\sin\theta \\ \sin\theta & \cos\theta \end{bmatrix}.$$

(2) Rotation about the x-axis can be represented by the matrix

$$M_{x-rot} = \begin{bmatrix} 1 & 0 \\ 0 & -1 \end{bmatrix}$$

so that if $z = (x,y)$ then $M_{x-rot}z = (x,-y)$.

(3) Rotation about the line through $(0,0)$ with slope angle θ is achieved by the multiplication of three matrices:

$$M = M_\theta M_{x-rot} M_{-\theta} = \begin{bmatrix} \cos(2\theta) & \sin(2\theta) \\ \sin(2\theta) & -\cos(2\theta) \end{bmatrix}.$$

(4) Rotation about the x-axis followed by a counter-clockwise rotation by θ is represented by the multiplication of two matrices $M = M_\theta M_{x-rot} = \begin{bmatrix} \cos\theta & -\sin\theta \\ \sin\theta & \cos\theta \end{bmatrix} \begin{bmatrix} 1 & 0 \\ 0 & -1 \end{bmatrix} = \begin{bmatrix} \cos\theta & \sin\theta \\ \sin\theta & -\cos\theta \end{bmatrix}$.

(5) For $p \in \mathbb{R}$, $pM = p \begin{bmatrix} a & c \\ b & d \end{bmatrix} = \begin{bmatrix} pa & pc \\ pb & pd \end{bmatrix}$.

(6) If $M = \begin{bmatrix} a & c \\ b & d \end{bmatrix}$ and $ad - bc \neq 0$, then the **inverse operator** is given by

$$M^{-1} = \begin{bmatrix} a & c \\ b & d \end{bmatrix}^{-1} = \frac{1}{ad - bc} \begin{bmatrix} d & -c \\ -b & a \end{bmatrix}.$$

Then $MM^{-1} = M^{-1}M = \begin{bmatrix} 1 & 0 \\ 0 & 1 \end{bmatrix}$, the identity matrix.

(7) Multiplying $z = (x, y)$ by a constant multiple of the identity matrix $\begin{bmatrix} p & 0 \\ 0 & p \end{bmatrix}$ will expand or contract z in Euclidean distance since $|(px, py)| = p\,|(x, y)|$, for $p \in \mathbb{R}$.

C.7 Homeomorphisms: Inverse maps on compact subsets of $\mathbb{R}^2$

Let f be a one-to-one map from a compact subset S of $\mathbb{R}^2$ onto a subset $T = f(S)$ of $\mathbb{R}^2$.

Definition C.22. The **inverse map** $g = f^{-1}$ of f is defined on the compact set T by the equation

$$g(f(s)) = s.$$

Note that, for $t \in T$,

$$f(g(t)) = t.$$

Proposition C.16. *Suppose f is a one-to-one, continuous map from a compact set $S \subset \mathbb{R}^2$ to $\mathbb{R}^2$. Then f^{-1} is continuous on the compact set $T = f(S)$.*

Proof. Suppose the sequence $\{s_i\} = \{g(t_i)\}$ does not converge. Let $\{s_{j_i}\}$ and $\{s_{k_i}\}$ be disjoint subsequences that converge to two points J and K in S. Then $t_{j_i} = f(s_{j_i}) \to f(J)$, $t_{k_i} = f(s_{k_i}) \to f(K)$ and $f(J) \neq f(K)$; so $\{t_i\}$ does not converge. □

Definition C.23. The map f is called a **homeomorphism** from S to T; T is said to be homeomorphic to S; S and T are said to be **homeomorphic**.

C.8 Problems

PROBLEM C.1. Use inequalities (C.4) to show that a sequence $\{z_k\} = \{(x_k, y_k)\}$ of points in $\mathbb{R}^2$ is a Cauchy sequence in $\mathbb{R}^2$ if and only if both $\{x_k\}$ and $\{y_k\}$ are Cauchy sequences in $\mathbb{R}$.

PROBLEM C.2. Show that a Cauchy sequence in $\mathbb{R}^2$ converges.

PROBLEM C.3. Suppose that a sequence $\{z_n\}$ of points in $\mathbb{R}^2$ converges to a point z. Let $\{z_{n_j}\}$ be a subsequence of $\{z_n\}$. Show that $\lim_j z_{n_j} = z$.

PROBLEM C.4. Let m be a map from $\mathbb{R}$ to $\mathbb{R}^2 : m(t) = (g(t), h(t))$, where each of g and h is a map from $\mathbb{R}$ to $\mathbb{R}$. Show that m is continuous at a point $t \in \mathbb{R}$ if and only if both g and h are continuous at t.

PROBLEM C.5. Let m be a map from $I = [0, 1]$ to $\mathbb{R}$. Let G be the associated graph map: $G(x) = (x, m(x))$, from $[0, 1]$ to $\mathbb{R}^2$.

(a) Show that m is continuous on I if and only if G is continuous on I.
(b) Show that if m is continuous on I, then the graph $G(I) = \{G(x) : x \in I\}$ is a compact subset of $\mathbb{R}^2$.

PROBLEM C.6. Let each of f and g be a continuous map from $\mathbb{R}^2$ to $\mathbb{R}^2$.

(a) Show that the map $f + g$ is a continuous map from $\mathbb{R}^2$ to $\mathbb{R}^2$.
(b) Show that the map fg is a continuous map from $\mathbb{R}^2$ to $\mathbb{R}^2$. The identity

$$\begin{aligned} & f(w)g(w) - f(z)g(z) \\ &= (f(w) - f(z))(g(w) - g(z)) + f(z)(g(w) - g(z)) + g(z)(f(w) - f(z)) \end{aligned}$$

is handy here.
(c) Show that if $g(z) \neq 0$, then the map $1/g$ is defined near z and is continuous at z.

PROBLEM C.7. Let $f : D \longrightarrow E$ and $g : E \longrightarrow F$ be continuous maps where D, E, and F are subsets of $\mathbb{R}^2$. Show that the composition $g \circ f$ of g with f defined by

$$(g \circ f)(z) = g(f(z))$$

is continuous on D.

PROBLEM C.8. Verify Proposition C.6.

PROBLEM C.9. Show that $\mathbb{R}$ is a connected set.

PROBLEM C.10. Show that for $r > 0$, $\{z \in \mathbb{R}^2 : |z| \leq r\}$ is a connected set. (Hint: Look at our proof of Proposition C.10.)

PROBLEM C.11. Show that $\mathbb{R}^2$ is a connected set.

PROBLEM C.12. Let f be continuous map from I to $\mathbb{R}$, and let $a \in \mathbb{R}$. Show that $\{x : f(x) \geq a\}$ is a closed subset of I.

PROBLEM C.13. Show that the map m used in the proof of Proposition C.15 is not continuous at w.

PROBLEM C.14. For $0 \leq x \leq 1$, put $f_k(x) = x^k$. Show that for each $x \in I = [0, 1]$ and each positive integer k, $f_k(x) \geq f_{k+1}(x) \geq 0$. Since bounded monotone sequences in $\mathbb{R}$ converge, the sequence $\{f_k(x)\}_{k \geq 1}$ converges for each $x \in I$. Put $f(1) = 1$ and $f(x) = 0$ if $0 \leq x < 1$. Show that the sequence $\{f_n\}$ converges monotonically downward to f on I. Note that f is not continuous on I.

PROBLEM C.15. Show that if a sequence $\{f_k\}$ of continuous maps on I to $\mathbb{R}$ converges pointwise and monotonically downward to a continuous map f on $[0, 1]$, then the sequence converges uniformly to f on $[0, 1]$. *Hints*: Put $g_n = f_n - f$; so, for each point $x \in I$, the sequence $\{g_n(x)\}$ converges monotonically downward to 0: the map m for which $m(I) = 0$. Use Problem C.12 and Proposition C.13 to show that the sequence $\{g_n\}$ converges to 0 uniformly on I as follows. Let $a > 0$; put $G_n = \{x : g_n(x) \geq a\}$ and show that there exists a positive integer n such that G_n is empty.

PROBLEM C.16. Give an example of a sequence $\{f_k\}$ of continuous maps on $[0, 1]$ to $\mathbb{R}$ that converges pointwise, but not uniformly, to a continuous map f on $[0, 1]$.

PROBLEM C.17. A non-empty subset S of $\mathbb{R}^2$ is said to be **disconnected** if there exist two non-empty, disjoint subsets A and B of $\mathbb{R}^2$ such that

- $A \cup B = S$,
- no point of B is a limit point of A: no sequence $\{b_k\}$ of points in B converges to a point $a \in A$, and
- no point of A is a limit point of B.

(a) Show that if S is the set of rational numbers in the unit interval $[0, 1]$, then S is a disconnected set.
(b) Show that if S is the set of rational numbers in the unit interval $[0, 1]$, then there do not exist two non-empty, disjoint closed subsets A and B of $\mathbb{R}^2$ such that $A \cup B = S$.
(c) Show that if S is a closed subset of $\mathbb{R}^2$ which is disconnected as defined above in this problem, then there exist two non-empty, disjoint, closed subsets A and B of $\mathbb{R}^2$ such that $A \cup B = S$.

PROBLEM C.18. Suppose that each of S and T is a closed subset of $\mathbb{R}^2$. Show that $S \cap T$ is a closed subset of $\mathbb{R}^2$. Suppose that $f : S \to \mathbb{R}^2$ and $g : T \to \mathbb{R}^2$ are continuous and that $f = g$ on $S \cap T$. Show that the common extension h of f and g ($h = f$ on S and $h = g$ on T) is continuous on $S \cup T$.

PROBLEM C.19. Define $f : \mathbb{N} \to \mathbb{R}$ by $f(2) = 2$ and $f(n) = 1$ if $n \neq 2$. Put $B = \{n + 1/n : n \in \mathbb{N}\}$. Define $g : B \to \mathbb{R}$ by $g(2) = 2$ and $g(n + 1/n) = 3$ if $n > 1$.

(a) Show that both $\mathbb{N}$ and B are closed sets.
(b) Show that any map $\varphi : \mathbb{N} \to \mathbb{R}$, is uniformly continuous.
(c) Show that any map $\psi : B \to \mathbb{R}$, is uniformly continuous.
(d) From the previous part both f and g are uniformly continuous. Because $f = g$ on $\mathbb{N} \cap B$, there is a common extension h of f and g to $\mathbb{N} \cup B$. Show that h is continuous, but **not** uniformly continuous, on $\mathbb{N} \cup B$.

PROBLEM C.20. Suppose that each of S and T is a closed subset of $\mathbb{R}^2$. Suppose that S is compact. Suppose that $f : S \to \mathbb{R}^2$ and $g : T \to \mathbb{R}^2$ are uniformly continuous. Suppose that $f = g$ on $S \cap T$. Show that the common extension h of f and g is uniformly continuous on $S \cup T$.

PROBLEM C.21. Suppose that $f : [0, \infty) \to \mathbb{R}$, is continuous. Suppose that there exists $c > 0$ such that f is uniformly continuous on $[c, \infty)$. Show that f is uniformly continuous on $[0, \infty)$.

PROBLEM C.22. Show that the map $z \to |z|$ is a continuous map on $\mathbb{R}^2$ to $\mathbb{R}^2$.

PROBLEM C.23. Let each of f and g be a continuous map from $\mathbb{R}^2$ to $\mathbb{R}^2$. Show that the map $g \circ f$:

$$(g \circ f)(z) = g(f(z)),$$

is continuous on $\mathbb{R}^2$.

PROBLEM C.24. Let f be a continuous map from $D \subset \mathbb{R}^2$ to $\mathbb{R}^2$. Show that the map $z \to |f(z)|$ is continuous on D.

PROBLEM C.25. Let each of f and g be a continuous map from $\mathbb{R}^2$ to $\mathbb{R}$. Define maps $f \vee g$ and $f \wedge g$ on $\mathbb{R}^2$ as follows:

$$(f \vee g)(z) = f(z) \vee g(z) = \max\{f(z), g(z)\}$$

and

$$(f \wedge g)(z) = f(z) \wedge g(z) = \min\{f(z), g(z)\}.$$

Show that

$$f \vee g = (1/2)(f + g) + (1/2)|f - g|,$$

and find the corresponding formula for $f \wedge g$. Explain why $f \vee g$ and $f \wedge g$ are continuous on $\mathbb{R}^2$.

PROBLEM C.26. Suppose that E is a compact subset of $\mathbb{R}$. Put

$$a = \inf\{y : y \in E\}$$

and

$$b = \sup\{y : y \in E\}.$$

(a) Show that a and b are in E and $E \subseteq [a, b]$.
(b) Suppose that $E \neq [a, b]$. Show that $[a, b]\backslash E$ is composed of a (perhaps finite) sequence of pairwise disjoint segments. These segments are called the *components (maximal, connected subsets)* of $[a, b]\backslash E$. For $x \in [a, b]\backslash E$, put

$$l_x = \sup\{y \in E : y < x\}$$

and

$$r_x = \inf\{y \in E : y > x\}.$$

Show that l_x and r_x are in E and the segment (l_x, r_x) is disjoint from E. Explain why

$$[a, b]\backslash E = \cup_{x \in (a,b)/E}(l_x, r_x).$$

Each set in the preceding union is a maximal, connected subset (a *component*) of $[a, b]\backslash E$; each of these components contains a rational number. There are only countably many rational numbers.
(c) Show that $len(E)$ is defined.

PROBLEM C.27. Suppose E_n is a convergent sequence of connected sets in $(\mathcal{K}, h)$ with limit set E. Show that E is connected. Hence, the connected compact sets comprise a closed subset of $(\mathcal{K}, h)$. (Peruse Propositions C.8 and C.11. You may wish to modify the proof of Proposition C.11.)

PROBLEM C.28. Show that for linear operators $M(N(z)) = MN(z)$.

PROBLEM C.29. Given linear operators M, N, and P show that $(MN)P = M(NP)$.

Appendix D

Infinite sets

This morsel defines finite, countably infinite, and uncountable sets and supplies information about infinite sets used in the text. In addition we hope that this material encourages you to learn more about infinite sets.

The set $\mathbb{N}$ of positive integers is composed of the least positive integer 1 and its successors: $\mathbb{N} = \{1\} \cup \{n+1 : n \in \mathbb{N}\}$. In Problem D.1 you can explore properties of the positive integers that are directly developed from this definition.

D.1 Countable and uncountable sets

Recall that a *sequence* $u = \{u_k\}_{k=1}^{\infty} = \{u_k\} = \{u(k)\} = \{(k, u(k)) : k \in \mathbb{N}\}$ is a function u with domain $\mathbb{N}$ and range $\{u(k) : k \in \mathbb{N}\}$. From a given a sequence $u = \{u_k\}$ we obtain a *related sequence* $v = \{v_n\} = \{u_{k_n}\}$ by the following process:

Set $k_1 = 1$. If

$$u_m \in \{u_k : k \leq k_n\} = \{v_j : j \leq n\}$$

for all $m > k_n$, then $k_{n+1} = k_n$; otherwise, k_{n+1} is the least integer m for which

$$u_m \notin \{v_j : j \leq n\}.$$

Example D.1. Suppose $u = \{u(k)\} = 1, 1, 2, 1, 2, 3, 1, 2, 3, 4, \ldots$. Recalling that $1 + 2 + \cdots + n = n(n+1)/2$, we have $v(n) = u(n(n+1)/2) = n$.

Example D.2. Suppose $u = 2, 0, 2, 0, \ldots$, $u(n) = 1 - (-1)^n$; then $v = 2, 0, 0, 0, \ldots$.

The related sequences u and v have the same range, and either v is a one-to-one map (as in Example D.1) or (as in Example D.2) there is a least positive integer j such that the first j entries in v are different and $v_i = v_j$ for all $i > j$.

Definition D.1. A set S is said to be **countable** if there is a sequence u whose range is S; S is said to be **countably infinite** if the sequence v related to u is a one-to-one map; otherwise S is said to be a **finite set with** j **elements**, where j is the least positive integer for which $v_i = v_j$ for all $i > j$. (Note that the first j entries in v are distinct.)

D.1.1 *The positive rational numbers are countably infinite*

To define an appropriate sequence u, we first identify a positive rational p/q with the ordered pair (p, q) and recall that $1 + 2 + \cdots + (n-1) = n(n-1)/2$. For $p + q = n + 1$, put $f(p,q) = n(n-1)/2 + q$. Then f is a one-to-one map with range $\mathbb{N}$. Now set $u(f(p,q)) = (p,q)$. That is,

$$\begin{aligned} u &= (1,1),(2,1),(1,2),(3,1),(2,2),(1,3),(4,1),(3,2),(2,3),(1,4),(5,1),\dots \\ &= \overleftrightarrow{1}, \overleftrightarrow{2}, \overleftrightarrow{1/2}, \overleftrightarrow{3}, 1, \overleftrightarrow{1/3}, \overleftrightarrow{4}, \overleftrightarrow{3/2}, \overleftrightarrow{2/3}, \overleftrightarrow{1/4}, \overleftrightarrow{5}, 4/2, 3/3, 2/4, \overleftrightarrow{1/5}, \overleftrightarrow{6}, \dots, \end{aligned}$$

where an arrow above an entry in u indicates that the entry is in the related sequence v:

$$\begin{aligned} &v(1) = u(1) = 1;\ v(2) = u(2) = 2;\ v(3) = u(3) = 1/2;\ v(4) = u(4) = 3; \\ &v(5) = u(6) = 1/3; \cdots. \end{aligned}$$

The sequence v related to u displays a one-to-one map from $\mathbb{N}$ onto the positive rationals. To define a one-to-one map of $w = \{w_k\}_{k=1}^{\infty}$ onto all the rationals, let $w_1 = 0$, $w_{2i} = v_i$, and $w_{2i+1} = -v_i$.

D.1.2 *The Cantor set is not a countable set*

To show that the Cantor set C is not a countable set, recall that each point in C has a unique base 3 representation in terms of 0's and 2's. Let $u = \{u_n\}$ be a sequence of points in C: $u_n = 0.p_{n,1}p_{n,2}p_{n,3}\cdots$, where $p_{n,j} \in \{0,2\}$. We can define a base 3 point $p = 0.p_1p_2p_3\cdots$ in C that is not in the range of u by setting $p_n = 2 - p_{n,n}$.

Problem D.3 demonstrates that C and I contain the *same* number of elements. Both of these sets have *more* elements than $\mathbb{N}$ because there is a map from I onto $\mathbb{N}$, but no map from $\mathbb{N}$ onto C or, equivalently, I.

D.1.3 *The continuum question*

Can you find a set that has more elements than $\mathbb{N}$ and fewer than I? (**Beware: You are not being given the continuum question as an exercise or a project!**) The *continuum hypothesis*, which suggests that the answer should be no, is discussed at great length in [Dauben (1970)] and elsewhere.

Having opened the door and peeked into the land of infinite sets, we depart.

D.2 Problems

PROBLEM D.1. Verify that the following four statements are equivalent.

(a) If a set S of positive integers satisfies A: $1 \in S$ and B: $n \in S \Rightarrow n+1 \in S$, then $S = N$.
(b) If a set S of positive integers satisfies A: $1 \in S$ and B: $k \in S$ when $k \leq n \Rightarrow n+1 \in S$, then $S = \mathbb{N}$.
(c) If a set S of positive integers satisfies A: $1 \in S$ and B: $k \in S$ when $k < n+1 \Rightarrow n+1 \in S$, then $S = \mathbb{N}$.
(d) Every non-empty subset of $\mathbb{N}$ contains a least positive integer.

PROBLEM D.2.

(a) Show that a countable union of countable sets is a countable set.
(b) Show that for $n \in \mathbb{N}$, the set $\{(j_1, j_2, \ldots, j_n) : j_i \in \mathbb{N}\}$ is a countable set.
(c) Show that the set of finite subsets of a countable set is a countable set.
(d) An algebraic number is a root of a polynomial with rational coefficients. Show that the set of algebraic numbers is countable.

PROBLEM D.3. Figure 1.14 displays a continuous map from the Cantor set C onto the unit interval I.

(a) Modify this map to obtain a one-to-one map from C onto I.
(b) Show that there is no one-to-one continuous map from C onto I. (Hint: Theorem C.3 and Proposition C.16 are available.)
(c) Show that there is a continuous map of the left half of the Cantor set onto I.
(d) Show that there is a map of C onto $\mathbb{R}$.
(e) Explain why there is no continuous map of C onto $\mathbb{R}$.
(f) Show that if you remove the point 1 from C, then there is a continuous map from this set onto $\mathbb{R}$.
(g) Display a map from C onto $\mathbb{R} \times \mathbb{R}$, the set of ordered pairs of points in $\mathbb{R}$.

Bibliography

Bailey, S., Kim, T. and Strichartz, R. (2002). Inside the Levy Dragon, *Am. Math. Monthly* **109**, pp. 689–703.

Barnsley, M. F. (1993). *Fractals everywhere*, 2nd edn. (Academic Press).

Crownover, R. M. (1995). *Introduction to fractals and chaos* (Jones and Bartlett).

Darst, R. B. (1993). The Hausdorff dimension of the nondifferentiability set of the Cantor function is $[\ln(2)/\ln(3)]^2$, *Proc. AMS* **119**, pp. 105–108.

Darst, R. B. (1995). Hausdorff dimension of non-differentiability points of Cantor functions, *Math. Proc. Camb. Phil.* **117**, pp. 185–191.

Darst, R. B., Palagallo, J. A. and Price, T. E. (1998). Fractal tilings in $\mathbb{R}^2$, *Mathematics Magazine* **71**, pp. 12–23.

Darst, R. B., Palagallo, J. A. and Price, T. E. (2008). Generalizations of the Koch curve, *Fractals* **16**, pp. 267–274.

Dauben, J. W. (1970). *Georg Cantor, His Mathematics and Philosophy of the Infinite* (Princeton University Press).

Drenning, S., Palagallo, J., Price, T. E. and Strichartz, R. S. (2005). Outer boundaries of self-similar tiles, *Experimental Mathematics* **14**, pp. 199–210.

Edgar, G. (1993). *Classics on fractals* (Addison Wesley).

Eidswick, J. A. (1974). A characterization of the nondifferentiability set of the Cantor function, *Proc. AMS* **42**, pp. 214–217.

Falconer, K. (1990). *Fractal geometry: mathematical foundations and applications* (John Wiley & Sons).

Hutchinson, J. E. (1981). Fractals and self-similarity, *Indiana Univ. Math. J.* **30**, pp. 713–747.

Keleti, T. (2006). When is the modified von Koch snowflake non-self-intersecting? *Fractals* **14**, pp. 245–249.

Mandelbrot, B. (1983). *The fractal geometry of nature* (Freeman).

Sagan, H. (1994). *Space filling curves* (Springer-Verlag).

Sullivan, D. (1983). *Conformal dynamical systems* (Springer Lecture Notes, Vol. 1007).

Weisstein, E. W. (2009a). Julia set, From MathWorld–A Wolfram Web Resource, `http://mathworld.wolfram.com/JuliaSet.html`.

Weisstein, E. W. (2009b). Mandelbrot set, From MathWorld–A Wolfram Web Resource, `http://mathworld.wolfram.com/MandelbrotSet.html`.

Solutions to selected problems

Problem 3.13(a) It suffices to find $M(1)$ because $M(2) = -M(1)$. $M(1)$ maps $(1, 0)$ onto $(a - 1/2, b)$ and (a, b) onto $(a/2 - 1/2, b/2)$. So

$$M(1) = \begin{bmatrix} p & r \\ q & s \end{bmatrix} \begin{bmatrix} 1 \\ 0 \end{bmatrix} = \begin{bmatrix} a - 1/2 \\ b \end{bmatrix},$$

and

$$\begin{bmatrix} a - 1/2 & r \\ b & s \end{bmatrix} \begin{bmatrix} a \\ b \end{bmatrix} = \begin{bmatrix} a^2 - a/2 + rb \\ ab + sb \end{bmatrix} = \begin{bmatrix} a/2 - 1/2 \\ b/2 \end{bmatrix}.$$

Thus, $s = 1/2 - a$ and $r = (-a^2 + a - 1/2)/b$ and

$$M(1) = \begin{bmatrix} a - 1/2 & (-a^2 + a - 1/2)/b \\ b & 1/2 - a \end{bmatrix}.$$

□

Problem 4.1. The proof is by induction. First observe that $f_{1/4}(0) = 0 \in S$ and $f_{1/4}(0.3) = .3 \in S$. Now suppose that for any $s = .t_1t_2t_3 \cdots t_n \in S$ $f_{1/4}(s) = s$ and consider

$$\begin{aligned} f_{1/4}(.3t_1t_2t_3 \cdots t_n) &= \frac{3}{4} + \frac{1}{4} f_{1/4}(s) \\ &= \frac{3}{4} + \frac{1}{4} s \\ &= .3t_1t_2t_3 \cdots t. \end{aligned}$$

The proof for the case $.0t_1t_2t_3 \cdots t_n$. is similar.

Problem 5.11. Let $A > 0$ satisfy

$$d_2(x, y) \leq A d_1(x, y).$$

Next, let $\varepsilon > 0$ and set $\delta = \varepsilon / A$. Then for any $x, y \in X$ satisfying

$$d_1(x, y) < \delta$$

we have

$$\begin{aligned} d_2\left(f\left(x\right), f\left(y\right)\right) &\leq Ad_1(f\left(x\right), f\left(y\right)) \\ &= Ad_1\left(x, y\right) \\ &< A\delta = \varepsilon. \end{aligned}$$

That is, f is continuous at any point $x \in X$ so it is continuous on X.

Problem 5.13. For a fixed real number x we can make $d(x,y)/d_1(x,y) = \left|x^2 + xy + y^2\right|$ as large as we wish by choosing y large.) We claim that $f : (\mathbb{R}, d_1) \longrightarrow (\mathbb{R}, d)$ is continuous. To see this let $\varepsilon > 0$, fix $x \in \mathbb{R}$, and set $\delta = \min\{1, \varepsilon/\left(1 + 2\left|x^3\right|\right)\}$. (Notice that δ depends on x in this case.) Now, if

$$d(x,y) = |x^3 - y^3| < \delta = \varepsilon/\left(1 + 2\left|x^3\right|\right)$$

then

$$\left|x^3 - y^3\right|\left(1 + 2\left|x^3\right|\right) < \varepsilon.$$

Because $1 > |x^3 - y^3| \geq \left|y^3\right| - \left|x^3\right|$ we have $\left|y^3\right| < 1 + \left|x^3\right|$ so

$$\left|x^3\right| + \left|y^3\right| < 1 + 2\left|x^3\right|.$$

Combining these inequalities we obtain

$$\epsilon > \left|x^3 - y^3\right|\left(\left|x^3\right| + \left|y^3\right|\right) \geq \left|\left(x^3 - y^3\right)\left(x^3 + y^3\right)\right| = \left|x^6 - y^6\right|.$$

Hence,

$$d(f\left(x\right), f\left(y\right)) = \left|\left(x^2\right)^3 - \left(y^2\right)^3\right| < \varepsilon$$

completing the proof of the claim since x was arbitrary.

Problem A.1. Choose $k \in \mathbb{N}$ such that if $n > k$, then $|y_n - y| < 1$. Put $M = \max\{|y_1|, |y_2|, \ldots, |y_k|, |y| + 1\}$. Then M is an upper bound for $\{|y_i| : i \in \mathbb{N}\}$.

Problem A.3. Put $T = \{-s : s \in S\}$ (reverse S). We know that multiplication by -1 reverses inequalities; so, if x is a lower bound for S, $-x$ is an upper bound for T. Thus, T is bounded above. We have shown that T has a least upper bound, call it L. Then, $G = -L$ is a lower bound of S. Moreover, if x is a lower bound for S, Then $-x \geq L$: $x \leq G$; so, G is the greatest lower bound of S.

Problem A.4(a) Let $\epsilon > 0$. Choose k such that $y_k < \inf S + \epsilon$. Then, for $j \in \mathbb{N}$,

$$\inf S \leq y_{k+j} \leq y_k < \inf S + \epsilon;$$

consequently, $|y_{k+j} - \inf S| < \epsilon$. $(0 \leq y_{k+j} - \inf S \leq y_k - \inf S < \epsilon.)$

Alternately, we can put $T = \{-s : s \in S\}$ (reverse S) and (as in our solution to exercise 1.3) use the fact that multiplication by -1 reverses inequalities to apply to corresponding result for non-decreasing sequences. □

Problem A.7. Because the sequence of intervals is nested, $a_k < a_{k+1} \leq b_{k+1} \leq b_k$. Thus, $\{a_k\}$ is non-decreasing and $\{b_k\}$ is non-increasing; $\{a_k\}$ is bounded above by b_1 and $\{b_k\}$ is bounded below by a_1. Since the sequence $\{a_k\}$ is non-decreasing and bounded, it converges to $a = \sup a_k$. Similarly, the non-increasing, bounded sequence $\{b_k\}$ converges to $b = \inf b_k$. Because $a_i < b_j$ for all positive integers i and j, $a = \sup a_k \leq \inf b_k = b$. To show that $\cap_{k\geq 1}[a_k, b_k] \supseteq [a, b]$, let $x \in [a, b]$. Then, for $k \geq 1$,

$$a_k \leq a \leq x \leq b \leq b_k;$$

thus, $x \in \cap_{k\geq 1}[a_k, b_k]$. To show that $\cap_{k\geq 1}[a_k, b_k] \subseteq [a, b]$, let $x \in \cap_{k\geq 1}[a_k, b_k]$. Then, for $k \geq 1$, $a_k \leq x \leq b_k$; thus,

$$a = \sup a_k \leq x \leq \inf b_k = b : x \in [a, b].$$

Problem A.9(a) Let $y_k \to y$. Let $\epsilon > 0$. Choose $n \in \mathbb{N}$ such that if $m \geq n$, then $|y_m - y| < \epsilon/2$. If $i \geq n$ and $j \geq n$, then

$$|y_i - y_j| \leq |y_i - y| + |y - y_j| < \epsilon/2 + \epsilon/2 = \epsilon.$$

Thus, $\{y_k\}$ is a Cauchy sequence. □

Problem A.9(b) Let $\epsilon > 0$ and choose a positive integer p such that if $k > p$, then $|y_k - y_p| < \epsilon/2$. For $m > p$ and $n > p$,

$$|y_m - y_n| \leq |y_m - y_p| + |y_p - y_n| < \epsilon/2 + \epsilon/2 < \epsilon.$$

□

Problem A.10. Let $\{y_{k_j}\}$ be a subsequence of $\{y_k\}$. Suppose $y_k \to y$. Let $\epsilon > 0$. Choose $m \in \mathbb{N}$ such that if $i \geq m$, then $|y_i - y| < \epsilon$. Since $k_j \geq j$, $|y_{k_j} - y| < \epsilon$ if $j \geq m$; thus, the subsequence $\{y_{k_j}\}$ converges to y.

Alternatively, according to Problem A.1, the sequence $\{y_k\}$ is bounded. Let $\{[a_k, b_k]\}$ be the associated sequence of intervals for the sequence $\{y_k\}$. Then $a_k \to a = y = b \leftarrow b_k$. Let $\{y_{k_j}\}$ be a subsequence of $\{y_k\}$. Let $\{[c_j, d_j]\}$ be the associated sequence of intervals for $\{y_{k_j}\}$. Since $\{y_{k_j}\}$ is a subsequence of $\{y_j\}$, referring to the first solution we have $a_{k_j} \leq c_j \leq d_j \leq b_{k_j}$. Thus, $c_j \to a = y = b \leftarrow d_j$; consequently, $\{y_{k_j}\}$ converges to y.

Problem A.15(a) We leave 1).⇒2)⇒3) for you to do. To verify 3)⇒1), we choose a sequence of epsilons that converge to 0: putting $\epsilon_n = 1/n$ will work fine. Let

$\{y_k\}$ be a sequence of points in $\mathbb{R}$ and let z be a number that satisfies statement (C).Let $k_1 = 1$. For $j > 1$, let k_j be the least positive integer that is greater than $k_{j-1}$and satisfies the inequality

$$\left|y_{k_j} - z\right| < \epsilon_j = 1/j.$$

The sequence $\{y_{k_j}\}$ converges to z. □

Problem A.15(b) Set $k_0 = 0$, and, for $j \geq 1$, let k_j be the least positive integer that is greater than k_{j-1} and satisfies the inequality $\left|y_{k_j} - u_j\right| < \epsilon_j = 1/j$. The sequence $\{y_{k_j}\}$ converges to u because

$$\left|y_{k_j} - u\right| \leq \left|y_{k_j} - u_j\right| + |u_j - u| \to 0.$$

□

Problem A.16(a) It suffices to choose a sequence that is composed of all the rational numbers in $[0, 1]$. □

Problem A.16(b) It suffices to choose a sequence that is composed of all the rational numbers. □

Problem A.16(c) It suffices to choose a sequence that is composed of all left end points in the construction of the standard Cantor set:

$$\overleftrightarrow{0, 2/3}, \overleftrightarrow{2/9, 2/3 + 2/9}, \overleftrightarrow{2/27, 2/9 + 2/27, 2/3 + 2/27, 2/3 + 2/9 + 2/27}, \ldots.$$

Block $n + 1$ is composed of the 2^n numbers obtained by adding $1/3^{n+1}$ to each number that is in the first n blocks. □

Problem A.16(d) Put $k_0 = 0$, and, for $j \geq 1$, let k_j be the least positive integer that is greater than k_{j-1} and satisfies the inequality $y_{k_j} > j$. □

Problem B.2. We know that $U - (U \cap V)$, $V - (U \cap V)$, and $U \cap V$ are three nonoverlapping intervals whose union is $U \cup V$. Hence, using equation B.3 the length of the interval $U \cup V$ is

$$\begin{aligned}\text{len}\,(U \cup V) &= \text{len}\,(U - (U \cap V)) + \text{len}\,(V - (U \cap V)) + \text{len}\,(U \cap V) \\ &= \text{len}\,(U) + \text{len}\,(V) - \text{len}\,(U \cap V)\end{aligned}$$

from which the result follows.

Problem B.3. Note that $[a, b] - (\cup_{1 \leq i \leq n} I_j)$ is the union of pairwise, nonoverlapping intervals, say,

$$[a, b] - (\cup_{1 \leq i \leq n} I_j) = \cup_{1 \leq k \leq m} C_k$$

so that

$$[a, b] = (\cup_{1 \leq i \leq n} I_j) \cup (\cup_{1 \leq k \leq m} C_k).$$

Then

$$\operatorname{len}([a,b]) = \sum_{1\le i\le n} \operatorname{len}(I_j) + \sum_{1\le k\le m} \operatorname{len}(C_k).$$

$\operatorname{len}\left([a,b] - \bigcup_{1\le i\le n} E_i\right) = (b-a) - \sum_{1\le i\le n} \operatorname{len}(E_i)$

Problem C.1. $\Rightarrow$ Let $\{z_k\} = \{(x_k, y_k)\}$ be a Cauchy sequence in $\mathbb{R}^2$. Let $\epsilon > 0$. Choose N such that if $j, k > N$ then $|z_j - z_k| < \epsilon$. According to inequalities (3.1), if $j, k > N$ then

$$|x_j - x_k| \le |z_j - z_k| < \epsilon$$

and

$$|y_j - y_k| \le |z_j - z_k| < \epsilon.$$

Consequently, both $\{x_k\}$ and $\{y_k\}$ are Cauchy sequences in $\mathbb{R}$.

$\Leftarrow$ Suppose $\{x_k\}$ and $\{y_k\}$ are Cauchy sequences in $\mathbb{R}$. Let $\epsilon > 0$. Choose N_x such that if $j, k > N_x$ then $|x_j - x_k| < \epsilon/2$. Choose N_y such that if $j, k > N_y$ then $|y_j - y_k| < \epsilon/2$. Let $N = \max\{N, N\}$. If $j, k > N$ then, according to inequalities (3.1),

$$|z_j - z_k| \le |x_j - x_k| + |y_j - y_k| < \epsilon/2 + \epsilon/2 = \epsilon.$$

Consequently, $\{z_k\}$ is a Cauchy sequence in $\mathbb{R}^2$.

Problem C.2. Let $\{z_k\} = \{(x_k, y_k)\}$ be a Cauchy sequence in $\mathbb{R}^2$. According to Problem C.1, both $\{x_k\}$ and $\{y_k\}$ are Cauchy sequences in $\mathbb{R}$. We know that Cauchy sequences in $\mathbb{R}$ converge. Thus, there exist x and y in $\mathbb{R}$ such that $x_k \to x$ and $y_k \to y$. By Proposition C.1, $z_k \to z$.

Alternate *Solution*: Show that $\{z_k\}$ is a bounded sequence in $\mathbb{R}^2$. According to Proposition C.2, the bounded sequence $\{z_k\}$ has a convergent subsequence $\{z_{k_j}\} : z_{k_j} \to z$. Let $\epsilon > 0$. Choose N_1 such that if $m, n > N_1$ then $|z_m - z_n| < \epsilon/2$. Choose N_2 such that if $j \ge N_2$ then $\left|z_{k_j} - z\right| < \epsilon/2$. Put $N = \max\{N_1, N_2\}$. Then for $j, n > N$,

$$|z_n - z| = \left|z_n - z_{k_j} + z_{k_j} - z\right| \le \left|z_n - z_{k_j}\right| + \left|z_{k_j} - z\right| < \epsilon/2 + \epsilon/2 = \epsilon.$$

Problem C.3. Suppose the sequence $\{z_k\}$ converges to z. Let $\{z_{k_j}\}$ be a subsequence of $\{z_k\}$. Let $\epsilon > 0$. Choose N such that if $j > N$ then $|z_j - z| < \varepsilon$. Because $k_j \ge j$, $\left|z_{k_j} - z\right| < \varepsilon$ if $j > N$. Thus, $z_{k_j} \to z$.

Problem C.5(a) $\Rightarrow$: Let $x \in I$ and let $\epsilon > 0$. Let $\delta > 0$ such that if $y \in I$ and $|y - x| < \delta < \epsilon/2$, then $|m(y) - m(x)| < \epsilon/2$. Consequently, if $y \in I$

and $|y-x|<\delta$, then $|G(y)-G(x)| = |(y,m(y))-(x,m(x))| \leq |y-x| + |m(y)-m(x)| < \epsilon/2+\epsilon/2 = \epsilon$.

$\Leftarrow$: Let $\epsilon > 0$. Let $\delta > 0$ such that if $y \in I$ and $|y-x| < \delta$, then $|G(y)-G(x)| < \epsilon$. Then for $y \in I$ with $|y-x|<\delta$ we have

$$|m(y)-m(x)| \leq |(y,m(y))-(x,m(x))| = |G(y)-G(x)| < \epsilon.$$

□

Problem C.5(b) Let m be continuous on I. To show that $G(I)$ is closed, let $\{G(x_k)\}$ be a Cauchy sequence in $G(I)$. We know that $\{G(x_k)\}$ converges to a point $w \in \mathbb{R}^2$; we will show that $w \in G(I)$. Because I is compact, the sequence $\{x_k\}$ has a convergent subsequence $\{x_{k_j}\}$; $x_{k_j} \to x \in I$. Because m is continuous on I, $m(x_{k_j}) \to m(x)$. Thus, $G(x_{k_j}) = (x_{k_j}, m(x_{k_j})) \to (x, m(x)) = G(x)$ because $x_{k_j} \to x$ and $m(x_{k_j}) \to m(x)$. Consequently, $w = G(x)$ because $G(x_{k_j}) \to w$.

To show that $G(I)$ is bounded, we suppose, on the contrary, that $G(I)$ is not bounded and arrive at a contradiction as follows. Let $\{G(x_k)\}$ be a sequence in $G(I)$. Suppose that $|G(x_k)| \to \infty$. Because $G(x_k) = (x_k, m(x_k))$ and $x_k \in I$, $|m(x_k)| \geq |(x_k, m(x_k))| - |x_k| \geq |G(x_k)| - 1 \to \infty$. As above, we have a subsequence $G(x_{k_j}) \to G(x)$. But, $\left|G(x_{k_j})\right| \to \infty$; we have a contradiction.

□

Problem C.6(a) Let $z \in \mathbb{R}^2$.Let $\epsilon > 0$. We need to find $\delta > 0$ such that if $|w-z| < \delta$, then $|(f(w)+g(w)) - (f(z)+g(z))| < \epsilon$. Because f is continuous at z, we can choose $\delta_f > 0$ such that if $|w-z| < \delta_f$, then $|f(w)-f(z)| < \epsilon/2$. Because g is continuous at z, we can choose $\delta_g > 0$ such that if $|w-z| < \delta_g$, then $|g(w)-g(z)| < \epsilon/2$. Let $\delta = \min\{\delta_f, \delta_g\}$. If $|w-z| < \delta$, then

$$\begin{aligned}
&|(f(w)+g(w)) - (f(z)+g(z))| \\
&= |(f(w)-f(z)) + (g(w)-g(z))| \\
&\leq |f(w)-f(z)| + |g(w)-g(z)| \\
&< \epsilon/2 + \epsilon/2 = \epsilon.
\end{aligned}$$

□

Problem C.6(b) Let $1 > \epsilon > 0$. Choose $\delta_f > 0$ such that if $|w-z| < \delta_f$, then

$$|f(w)-f(z)| < \frac{\epsilon}{3(|g(z)|+1)}.$$

Choose $\delta_g > 0$ such that if $|w-z| < \delta_g$, then

$$|g(w)-g(z)| < \frac{\epsilon}{3(|f(z)|+1)}.$$

Put $\delta = \min\{\delta_f, \delta_g\}$. Then $|w - z| < \delta$ implies (check the details)

$$
\begin{aligned}
&|f(w)g(w) - f(z)g(z)| \\
&\leq |(f(w) - f(z))(g(w) - g(z))| + |f(z)(g(w) - g(z))| + |g(z)(f(w) - f(z))| \\
&< \frac{\epsilon}{3}\frac{\epsilon}{3} + \frac{\epsilon}{3} + \frac{\epsilon}{3} < \epsilon.
\end{aligned}
$$

□

Problem C.6(c) Let $z \in \mathbb{R}^2$ and $\epsilon > 0$. Choose $\delta > 0$ such that if $|w - z| < \delta$, then

$$|g(w) - g(z)| < \min\left\{\frac{\epsilon\,|g(z)|^2}{2}, \frac{|g(z)|}{2}\right\}.$$

For $|w - z| < \delta$, (check the details)

$$|g(w)| > |g(z)/2|$$

and

$$\left|\frac{1}{g(w)} - \frac{1}{g(z)}\right| = \left|\frac{g(z) - g(w)}{g(z)g(w)}\right| < \left|\frac{g(z) - g(w)}{g(z)g(z)/2}\right| < \epsilon.$$

□

Problem C.8. First observe that a closed and bounded set S is compact: a sequence x_i in S has a convergent subsequence x_{i_j} because S is bounded; the limit of the subsequence x_{i_j} is in S because S is closed. To show that a compact subset of $\mathbb{R}^2$ must be closed and bounded, let A be a subset of $\mathbb{R}^2$. Suppose that A is not closed; let $\{a_k\}$ be a sequence of points in A that converge to a point b which is not in A. No subsequence $\{a_{k_j}\}$ of the sequence $\{a_k\}$ can converge to a point in A because $\{a_{k_j}\}$ converges to b. Therefore, if A is not closed, then A is not compact. Suppose that A is not bounded; let $\{a_k\}$ be a sequence of points in A with $|a_k| > k$: $|a_k| \to \infty$. No subsequence of the sequence $\{a_k\}$ can converge. Therefore, if A is not bounded, then A is not compact.

Problem C.9. Suppose that $\mathbb{R}$ is disconnected. Then $\mathbb{R} = A \cup B$, where A and B are non-empty, closed and disjoint. Let $a \in A$ and $b \in B$. Choose $n \in \mathbb{N}$ such that both a and b are in $S = [-n, n]$. The sets $A \cap S$ and $B \cap S$ are non-empty, closed and disjoint. However,

$$(A \cap S) \cup (B \cap S) = (A \cup B) \cap S = \mathbb{R} \cap S = S.$$

Consequently, we have a contradiction because we know that $S = [-n, n]$ is connected.

Problem C.18. Let $\epsilon > 0$. $S \cup T$ is composed of three disjoint parts: $S \cap T$, $S \backslash T$ and $T \backslash S$. We consider each of the three parts separately.

(1) Let $x \in S \cap T$. Choose $\beta > 0$ such that if $y \in S$ and $|y - x| < \beta$, then $|f(y) - h(x)| < \epsilon$. Choose $\gamma > 0$ such that if $y \in T$ and $|y - x| < \gamma$, then $|g(y) - h(x)| < \epsilon$. Let $\delta = \min\{\beta, \gamma\}$. If $y \in S \cup T$ and $|y - x| < \delta$, then $|h(y) - h(x)| < \epsilon$.
(2) Let $x \in S/T$. Because T is closed, $d(x, T) = inf\{|x - t| : t \in T\} > 0$. Choose $\beta > 0$ such that if $y \in S$ and $|y - x| < \beta$, then $|f(y) - f(x)| < \epsilon$. Put $\delta = \min\{\beta, d(x, T)\}$. Then $|h(y) - h(x)| < \epsilon$ if $|y - x| < \delta$.
(3) Let $x \in T \backslash S$. Because S is closed, $d(x, S) > 0$. Choose $\beta > 0$ such that if $y \in T$ and $|y - x| < \beta$, then $|g(y) - g(x)| < \epsilon$. Put $\delta = \min\{\beta, d(x, S)\}$. Then $|h(y) - h(x)| < \epsilon$ if $|y - x| < \delta$.

Problem C.19(c) To show that both φ and ψ are uniformly continuous, put $\delta = 1/2$: If x and y are in $\mathbb{N}$ and $|x - y| < 1/2$, then $x = y$; the same statement applies to B. □

Problem C.19(d) Because both $\mathbb{N}$ and B are closed sets and $f = g$ on $\mathbb{N} \cap B = \{2\}$, Problem C.18 allows us to assert that the common extension h of f and g is continuous on $\mathbb{N} \cup B$. Now we verify that h is not uniformly continuous. Choose $\epsilon \in (0, 2)$. Let $\delta > 0$. If $1/n < \delta$, then $|(n + 1/n) - n| < \delta$ and $|h(n + 1/n) - h(n)| = 2 > \epsilon$. □

Problem C.20. We know (Problem C.18) that h is continuous on $S \cup T$. Let $\epsilon > 0$. Choose $\delta_s > 0$ such that if u and v are in S and $|u - v| < \delta_s$, then

$$|f(u) - f(v)| < \epsilon/2.$$

Choose $\delta_t > 0$ such that if u and v are in T and $|u - v| < \delta_t$, then

$$|f(u) - f(v)| < \epsilon/2.$$

Let $\delta = \min\{\delta_s, \delta_t\}$ Suppose there exists a sequence $\{(x_n, y_n)\}$ in $S \cup T$ such that $|x_n - y_n| < 1/n$ and $|f(x_n) - f(y_n)| \geq \epsilon$. Since both f and g are uniformly continuous, there exists $N \in \mathbb{N}$ such that if $n \geq N$, then exactly one of x_n and y_n is in S, and the other is in T. Without loss of generality, we suppose that $x_n \in S$ and $y_n \in T$ for $n \geq N$. Because S is compact, we can choose a convergent subsequence $\{x_{n_k}\}$ of $\{x_n\}_{n \geq N}$: $x_{n_k} \to x \in S$. Because $|x_n - y_n| < 1/n$, $y_{n_k} \to x$ too. Consequently (T is closed and $y_n \in T$ for $n \geq N$), $x \in S \cap T$. Thus, for $1/n_k < \delta$, we have a contradiction:

$$|f(x_{n_k}) - f(x)| < \epsilon/2, \; |f(y_{n_k}) - f(x)| < \epsilon/2, \text{ and } |f(x_{n_k}) - f(y_{n_k})| \geq \epsilon.$$

Problem C.21. Because f is continuous on $[0, c]$ and $[0, c]$ is compact, f is uniformly continuous on $[0, c]$. Put $S = [0, c]$, $T = [c, \infty)$, and apply Problem C.20.

Problem C.22. Applying the triangle inequality we have

$$|z| = |(z - w) + w| \leq |z - w| + |w|$$

so that

$$|z| - |w| \leq |z - w| .$$

Likewise,

$$|w| - |z| \leq |w - z| .$$

Combine these results to get

$$- |z - w| \leq |z| - |w| \leq |z - w| ,$$

or

$$||z| - |w|| \leq |z - w| .$$

This is a very handy inequality. Explain it geometrically in $\mathbb{R}^2$.

Index